Bella Bathurst was born in 1969. She is a freelance journalist and lives in London.

For automatic updates on Bella Bathurst visit harperperennial.co.uk and register for AuthorTracker.

From the reviews of *The Lighthouse Stevensons*:

'An inspiring account of men pushed to the limits and beyond on offshore slivers of black rock exposed to sea and gales; of masons, carpenters, blacksmiths and storemen lodged on site in elevated and cramped capsules that could become prisons for days on end when storms struck'

DAVID CAMERON, *Daily Telegraph*

'Bella Bathurst has uncovered a fascinating story. From the smugness and safety of our own times, it is no bad thing to have illuminated such giants'

PHILIP MARSDEN, *Mail on Sunday*

'"All that stone and history and effort, you think, just for a lightbulb". A gripping history, beautifully written'

BRIAN CASE, *Time Out*

'Bathurst's engaging story of three generations of the Stevenson family is not simply a tale of the Scottish Enlightenment. It's not just the story of that time when its engineers and scientists first mastered our relationship with a still savage nature. It is also the narrative of a remarkable dynasty and its struggle with commerce and art' RICHARD COOK, *Guardian*

'An enthralling story, vivaciously recounted'

ALAN TAYLOR, *Observer*

'A fine and scrupulous book . . . Robert Louis Stevenson would have relished the tale told here' IAN BELL, *Scotsman*

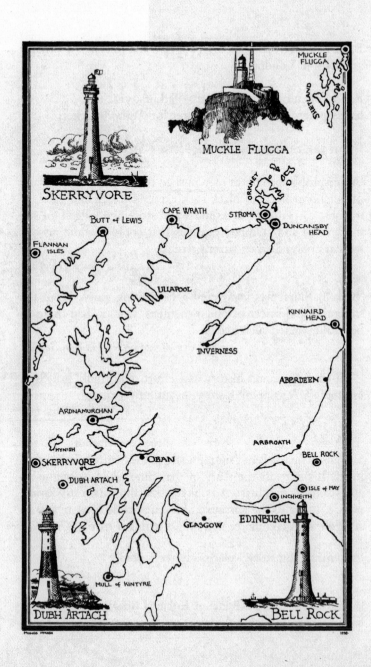

SKERRYVORE

MUCKLE FLUGGA

MUCKLE FLUGGA

SHETLAND

ORKNEY

CAPE WRATH

STROMA

DUNCANSBY HEAD

Butt of LEWIS

FLANNAN ISLES

ULLAPOOL

KINNAIRD HEAD

INVERNESS

ABERDEEN

ARDNAMURCHAN

HYNISH

ARBROATH

BELL ROCK

SKERRYVORE

OBAN

DUBH ARTACH

ISLE of MAY

INCHKEITH

EDINBURGH

GLASGOW

DUBH ARTACH

MULL of KINTYRE

BELL ROCK

MUNGO MCCOSH

THE
LIGHTHOUSE
STEVENSONS

*The extraordinary story of
the building of the Scottish lighthouses
by the ancestors of
Robert Louis Stevenson*

BELLA BATHURST

HARPER PERENNIAL

To my mother,
and to Lucy and Flora.

Harper Perennial
An Imprint of HarperCollins*Publishers*
77–85 Fulham Palace Road
Hammersmith
London w6 8jb

www.harperperennial.co.uk

This edition published by Harper Perennial 2005
5

Previously published in paperback by Flamingo 2000

First published in Great Britain by HarperCollins*Publishers* 1999

A catalogue record for this book
is available from the British Library

ISBN 978-0-00-720443-4

Frontispiece map by Mungo McCosh

Set in Janson with Spectrum display

Printed and bound in Great Britain by
Clays Ltd, St Ives plc

CONTENTS

There are spaces still to be filled
before the map is completed –
though these days it's only
in the explored territories
that men write, sadly,
Here live monsters.

NORMAN McCAIG, *Old Maps and New*

PREFACE

This is not – nor was it ever intended to be – a definitive biography of the Lighthouse Stevensons. When I began research in January 1996, I realised that any attempt to write a comprehensive biography of all four generations of the Stevensons would be both futile and, given the range and technicality of their work, probably quite wearing as well. And yet to write the history of just one of their lives would be to leave an incomplete picture. The Stevensons were, in the best and worst of senses, a family business and are perhaps most easily understood in that context. The solution I came to was to concentrate on the time between 1786 and 1890 when the first four Lighthouse Stevensons were working around the Scottish coastline, and to focus in detail on those four lights that were most closely associated with their respective engineers. The result of this selection is not quite biography and not quite history. If my selections at times seem arbitrary or incomplete, I can only apologise. The life and works of Robert Louis Stevenson have been the subject of innumerable biographies, studies, critiques and analyses, not to mention his own autobiographical writings. The story of his ancestors, by comparison, remains a relatively unworn path. I would hope therefore that this book would be seen as a kind of taster for the subject, and that anyone wanting to search further would be able to do so. There is more, much more, in the lives and works of the Lighthouse Stevensons than any one book could ever hope to encompass.

ACKNOWLEDGEMENTS

Since 1786, all the Northern Lights have been administered by the Northern Lighthouse Board. In March 1998 the last of the lights was automated but the loss of the keepers has if anything meant an increase in the NLB's workload. At present, they are responsible for 84 'major' lights (those originally manned), 112 minor lights (most of which were always automatic), and a network of beacons, buoys and radar beacons around the coast. They are also still building lighthouses. The lights at Haskeir, Gasker and Monach Isles have all been constructed in the last few years to serve the oil tankers on their route west of Lewis to the terminal at Sullom Voe on Shetland. Whatever the arguments over the need for the lights, it is evident that the NLB's role and purpose is as powerful now as when it was first established.

This book could not have been written without their assistance. As well as plundering their archive, their picture library and their stationery cupboard, I was also given generous access to the remaining keepers and the lights themselves. Lorna Grieve, the NLB's Information Officer, answered questions, organised visits and helped trace material; I owe her heartfelt thanks and a ream of photocopying paper. Bill Paterson read and corrected the first draft, while Chief Executive James Taylor offered help and encouragement both for the book and for several articles during 1997. In London, Jane Wilson offered access to the Trinity House archive and helped with queries. I am also hugely grateful to the crew of the lighthouse ship *Pharos* and the keepers at Cape Wrath, Fair Isle, Butt of Lewis, Rinns of Islay, Duncansby Head and Mull of Kintyre for making me welcome and answering questions.

Quentin Stevenson and Jean Leslie also gave me help with material and with building an image of their ancestors. Simon Leslie, great great great grandson of Robert Stevenson, fulfilled the role of lawyer, friend and Stevenson with his usual style and discretion. I am also grateful to Hector McPhail, Pat Lorimer, Craig Mair, the Multiple Sclerosis Society, HM Coastguard, the Royal Yachting Association, the RNLI and the National Libraries of Scotland for information and assistance at various stages. Thanks also to Pete Jinks for his helpful suggestions for titles (which I helpfully ignored), Alan Taylor at the *Scotsman* for ideas and a reference, Alex and Danny Renton for answering endless questions about the sea and sailing, Euan Ferguson for comments and incentives, Ruaridh Nicoll for help with editing, Alexa de Ferranti for encouragement and kindness, Sandy Milne for inspiration and Victoria Hobbs and Michael Fishwick for turning this from an idea into a book.

LIST OF ILLUSTRATIONS

The Isle of May, Scotland's first lighthouse.
Winstanley's Eddystone lighthouse.
Smeaton's Eddystone lighthouse.
James Rennie's design for the Bell Rock.
Robert Stevenson's design for the Bell Rock light.
Cross-section of the Bell Rock.
Robert Stevenson, 1814.
The Bell Rock light room.
Robert Stevenson in later life, *courtesy Stevenson Family Collection*.
J.M.W. Turner's illustration of the Bell Rock during a storm.
The Eddystone, Skerryvore and Bell Rock lights.
Elevation of Alan Stevenson's Skerryvore light.
The only known portrait of Alan Stevenson, *courtesy Stevenson Family Collection*.
Cross-section of Skerryvore.
The temporary barracks at Skerryvore.
Alan's design for the Ardnamurchan.
The impossible lighthouse, Muckle Flugga.
David Stevenson *courtesy Stevenson Family Collection*.
Thomas Stevenson © *Scottish National Portrait Gallery*.
Robert Louis Stevenson © *Scottish National Portrait Gallery*.
The light at Dhu Heartach.
Raising coal to the tower at Dhu Heartach.
The barracks on Dhu Heartach.
Dhu Heartach.

THE LIGHTHOUSE STEVENSONS

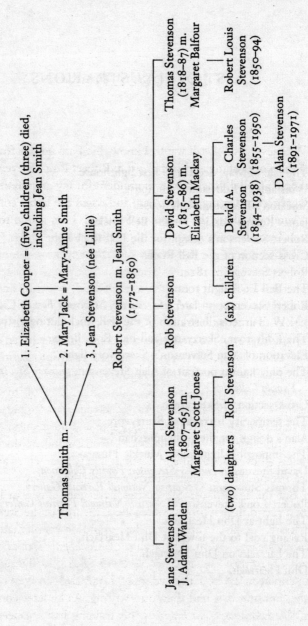

Thomas Smith m.

1. Elizabeth Couper = (five) children (three) died, including Jean Smith

2. Mary Jack = Mary-Anne Smith

3. Jean Stevenson (née Lillie)

Robert Stevenson m. Jean Smith
(1772–1850)

Jane Stevenson m.
J. Adam Warden

Alan Stevenson
(1807–65) m.
Margaret Scott Jones

Rob Stevenson

(two) daughters

Rob Stevenson

(six) children

David Stevenson
(1815–86) m.
Elizabeth Mackay

David A.
Stevenson
(1854–1938)

Charles
Stevenson
(1855–1950)

D. Alan Stevenson
(1891–1971)

Thomas Stevenson
(1818–87) m.
Margaret Balfour

Robert Louis
Stevenson
(1850–94)

INTRODUCTION

'Whenever I smell salt water, I know that I am not far from one of the works of my ancestors,' wrote Robert Louis Stevenson in 1880. 'The Bell Rock stands monument for my grandfather, the Skerry Vhor for my Uncle Alan; and when the lights come out at sundown along the shores of Scotland, I am proud to think they burn more brightly for the genius of my father.' Louis might have been the most famous of the Stevensons, but he was not the most productive. Between 1790 and 1940, eight members of the Stevenson family planned, designed and constructed the ninety-seven manned lighthouses that still speckle the Scottish coast, working in conditions and places that would be daunting even for modern engineers. The same driven energy that Louis put into writing, his ancestors put into lighting the darkness of the seas. The Lighthouse Stevensons, as they became known, were also responsible for a slew of inventions in both construction and optics and for an extraordinary series of developments in architecture, design and mechanics. As well as lighthouses, they built harbours, roads, railways, docks and canals all over Scotland and beyond. They, as much as anyone, are responsible for their country's appearance today.

But the Lighthouse Stevensons have gone down in history for a very different profession. Robert, the first of the Stevenson dynasty, despised literature; his grandson perpetuated his family's name with it. The author of *Kidnapped*, *Treasure Island* and *The Strange Case of Dr Jekyll & Mr Hyde* initially trained as an engineer. To his father's dismay, Louis escaped aged twenty-one, first into law and then into writing. As he later confessed in *The Education of an Engineer*, his training had not been used in quite the way his father intended. 'What I gleaned, I am sure I do not know; but in deed I had already my own private determination to be an author; I loved the art of words and the appearances of life; and *travellers*, and *headers*, and *rubble*, and *polished ashlar*, and *pierres perdues*, and even the thrilling question of the *string course*, interested me only (if they interested me at all) as properties for some possible romance or as words to add to my vocabulary.'

With age and distance, Louis recovered pride and affection in the Stevenson trade. He wrote with awe of his grandfather's work on the Bell Rock lighthouse, and of his father's melancholic genius for design and experimentation. He wrote about almost every aspect of his own brief and unhappy time as an apprentice, in essays, letters, introductions and memoirs. Most importantly, Louis alchemised his experiences around the ragged coasts of the north into the gold of his best fiction. *Treasure Island* and *Kidnapped* both contain salvaged traces of his early career. The further he grew away from engineering, the more he felt towards it. He was sea-marked, and he knew it. He also recognised, with some discomfiture, that his own fame was swallowing up the recognition his family deserved. In 1886, far from Edinburgh, he wrote crossly to his American publishers,

> My father is not an 'inspector' of lighthouses; he, two of my uncles, my grandfather, and my great grandfather in succession, have been engineers to the Scotch Lighthouse service; all the sea lights in Scotland are signed with our name; and my father's services to lighthouse optics have been distinguished indeed. I might write books till 1900 and not serve humanity so well; and it moves me to a certain impatience, to see the little, frothy bubble that attends the author his son, and compare it with the obscurity in which that better man finds his reward.

Louis was being only a little disingenuous. He liked recognition and, to an extent, courted it. But his plaintive belief that his family deserved the same acknowledgement seems far-sighted now. Even at the height of the Victorian engineering boom, great men went unnoticed and exceptional feats unacknowledged. Louis did his best to remedy the injustice, but also recognised that the Stevensons hardly helped themselves. Not one member of the family ever took out a patent on any of their inventions in design, optics or architecture. All of them believed that their works were for the benefit of the nation as a whole and therefore unworthy of private gain. They were only engineers, after all; they worked to order or conscience, and were only rarely disposed to flightier moments of reflection. What pride they had in their creations they put down to the advantages of forward planning and the benevolence of the Almighty. And Louis, the tricky, charming black sheep of the family, stole all the fame that posterity had to give.

Two hundred years ago, when the first lights were built around the Scottish coast, no one talked much of security. The first beacons for mariners were either coal fires or high coastal towers in which candles burned through the night. Only a few of these fires were ever constructed in Scotland, since they consumed fuel at a voracious pace and were usually extinguished by bad weather. Thus by the mid-eighteenth century, the Scottish coast had become notorious for shipwrecks. In 1799, seventy vessels were lost in the Firth of Tay alone. Along with the physical dangers, there were also the human ones. Bands of wreckers would lie in wait for beached ships, hoping for chances of loot. Something, it was becoming obvious, would have to be done.

Those engineers who did come forward were more like pioneers than bureaucrats. To place a building on a rock in the Atlantic ocean was, after all, a formidable endeavour. The pressures of wind, wave, tide and weather on a lighthouse were exceptional. No other building, not even a harbour, had to have quite the same mixture of tenacity and flexibility as the sea-towers did. Any construction in mid-ocean had to be capable of resisting waves which, when roused, could hurl several hundred tons of water at anything in their way. Every one of the rock lighthouses in Scotland was built with stone walls at least nine feet thick at the base; anything less, and they would not have lasted the first gale of winter. To build something under such pressures at a time when the only materials available were stone, wood, glass and metal was nothing short of miraculous. There was no concrete, or cranes, or hydraulic lifting equipment; there were no helicopters or pneumatic drills. Dynamite was a new and fickle builder's tool to be treated with extreme caution. Mortar was strong but unpredictable, requiring expert mixing and split-second timing. Haulage, in many cases, was provided by horses, who did not take kindly to precipitous cliffs and needed as much tending as any of the workmen. Equipment and materials were brought by sailboat, which ran exactly the same risks as any other ship. As the early engineers discovered, building 130-foot pillars in the middle of a hostile ocean required skills and tools that had not yet been invented. As often as not, they had to make it up as they went along.

From its slow beginnings, the organisation of lights was divided by nation: the English, the Scots and the Irish all had, and still have, separate administrations. The English service, which was founded in 1514 as the Most Glorious and Undivided Trinity of St Clement in the Parish of Deptford Strond in the County of Kent (later foreshortened to Trinity House), developed in piecemeal fashion. For a period of 300 years or so, most of its lights were built and maintained by individuals who had been granted charters. Although it did mean that the most hazardous parts of the English coast were lit, the lights' construction was erratic and their maintenance wayward. Pepys, who was Master of Trinity House from 1676 to 1689, found private charters disgraceful. While still at the Admiralty, he wrote critically of 'the evil of having lights raised for the profit of private men, not for the good of the public seamen, their widows and orphans'. In theory, the private owners could build, light and staff the towers in any way they chose in return for a small annual rent. Several were taken over by Trinity House once the lease had expired or had become suitably profitable. By 1800, the combination of extortionate private dues and inconsistent public ones was causing uproar among shipowners plying the English coastline. The situation had, in effect, become a form of legalised extortion: by 1818, Trinity House was reaping an annual profit of around £50,000. The few attempts to reform the situation usually resulted in an undignified tussle between Trinity House, the Crown and the landowners for a portion of the money. It was not until 1836 that Trinity House bought back all the lighthouses in private hands, an undertaking which cost them over £1 million.

The Scottish lights, by contrast, were almost all built within the space of a century. The archetypal lighthouse on its lonely rock in a lonely sea is largely the product of a Scottish imagination and a Scottish sense of endeavour. Augustin Fresnel's nineteenth-century refinements to oil lighting and high-powered lenses were matched by equally significant developments in lighthouse construction and marine technology. The Stevenson family took up the challenge of their times, blended it with the scientific breakthroughs of their day and brought both to a point of near-perfection. As Louis later noted, engineering 'was not a science then. It was a living art, and it visibly grew under the eyes and between the hands of its practitioners.'

In 1786 Louis's grandfather, Robert, went into partnership with his stepfather Thomas Smith, then the Engineer to the Board of Northern Lighthouses. The two began replacing the flickering and unreliable fires first with oil lamps and later with a system of fixed lights using either gas or oil. In 1807, Robert started work on the Bell Rock, a vicious granite reef off the coast of Arbroath. The reef was submerged at high tide and only partially exposed at low tide, so Robert and his staff were forced to play a nervous game of waiting with the sea and the weather. Lighthouses were not Robert's only preoccupation, however. 'Scotland itself,' as a biographer of his grandson later remarked, 'was his drawing board.' He was also responsible for the construction of the east side of Edinburgh, driving a route from Prince's Street to the Calton Hill and constructing the Georgian order of Waterloo Place and Regent Road. Robert was also a man of a most particular mood and time, a self-made bootstrap businessman who, like many of his engineering contemporaries, used the opportunities of the post-Enlightenment world to haul himself out of poverty and into society. He placed his faith in improvement and industry, and remained to the end a devout believer in the most conservative of virtues. He was, according to Louis, 'a man of the most zealous industry, greedy of occupation, greedy of knowledge, a stern husband of time, a reader, a writer, unflagging in his task of self-improvement.' Louis, it would seem, was not the only person to be daunted by him.

Robert had three sons who became engineers, Alan, David and Thomas. Alan was a classical scholar, musical, gifted and noted for his early championing of Wordsworth. Nevertheless, he suppressed the artistic side of himself to go into the family firm, becoming, like his father before him, the Commissioner of the Board of Northern Lights. He is remembered as a shrewd and brilliant engineer, whose greatest professional triumph was the construction of Skerryvore lighthouse on a ragged clump of rocks twelve miles west of Tiree. As Sir Walter Scott noted when he visited the site, 'It will be a most desolate position for a lighthouse – the Bell Rock and Eddystone a joke to it.' The light took five years to build, and, despite a fire in the 1950s, still stands today. Louis considered it 'the noblest of all extant deep-sea lights'.

David Stevenson took up Alan's position on his retirement.

His greatest achievement was the construction of the light at Muckle Flugga, the most northerly of all the Scottish lighthouses. Constructed as a temporary light to aid British naval convoys on their way to the Crimea, it was placed on the summit of a wave-washed miniature Matterhorn. Westminster insisted that the light should be working within six months; David's exceptional skill as an engineer ensured that, even with winter seas crashing 200 feet over the rock, it was finished in time.

Thomas Stevenson, Louis's father, was responsible for the construction of twenty-seven on-shore and twenty-five off-shore lighthouses. He built the light at Dhu Heartach, an isolated mass of rock off the coast of Mull, which Louis later used as source material for *Kidnapped*. At one point during its construction, fourteen men were trapped for five days in a temporary barracks while a ferocious gale pounded the rock. Louis records the foreman desperately playing his fiddle to quell the sound of the sea's rage. Thomas is also remembered for having taken Fresnel's optical developments several stages further, harnessing the strange new science of electricity and building a series of revolving lights in a bright and sturdy circuit which finally enclosed the whole of Scotland. Thomas was 'a man of somewhat antique strain', recalled Louis after his death, 'with a blended sternness and softness that was wholly Scottish and at first somewhat bewildering; with a profound essential melancholy of disposition and (what often accompanies it) the most humorous geniality in company; shrewd and childish; passionately attached, passionately prejudiced; a man of many extremes, many faults of temper, and no very stable foothold for himself among life's troubles.' For much of the time, he was working in conjunction with David. The two brothers complemented each other well; David tended towards details, Thomas towards inventiveness. Both were also preoccupied with more prosaic matters, including the establishment of precise and reliable systems for constructing, surveying, lighting, supplying and staffing their fast-growing constellation of lights. Thomas's weakest point, of course, was Louis, his only child, sickly, contrary and a source of constant worry. Thomas's wistful attempts to entice Louis into the family firm were fruitless. The rest is literary history.

David had two sons, David A. and Charles. Both are mainly now known for their refinements to the existing systems and their shrewd maintenance of the NLB. Most of the gaudier feats of engineering had been completed by Robert, Alan, David and Thomas. There seemed little for the grandchildren to do but tinker with what already existed. But the Stevensons' achievements in engineering, science and optics have lasted far beyond their lifespan. Anyone who has ever travelled the coast of Scotland has probably had cause to thank them. Their lights, built into the rocks of the most inhospitable land in Britain, have gone on shining for almost two centuries. Listen to the Shipping News today, and you are listening to part of the legacy of the Lighthouse Stevensons.

ONE

Yarmouth

Captain George Manby had reached the age of forty without having contributed significantly to life. His childhood in Yarmouth had been undistinguished, his military career nondescript, and by early middle age he had sunk deeply into debt. Apart from an incident in 1800 when he appeared wild-eyed at the Secretary at War's door offering to assassinate Napoleon – an offer which the Secretary politely declined – Manby seemed an unlikely candidate for immortality. His naval colleagues also noted cynically that the only battle scar he had yet earned was a gunshot wound, allegedly sustained while running away from a duel.

The death of Nelson during the Battle of Trafalgar in 1805 changed all that. Manby had been at school with Nelson, and although the two had not been friends, Manby still regarded the admiral with affection. When Nelson died, Manby was spurred into action. Inspired by his hero's example and impressed by the public grief over his loss, Manby concluded that his best chance of fame lay in saving lives; in particular, saving lives at sea. It was a startling choice. Manby's only marine experience until then had been an unsuccessful spell as a naval lay captain on a frigate heading for Dublin. The ship had foundered off the Irish coast, and, once embedded upon the lee shore, began sinking fast. Manby wrote later that 'the striking of the ship was the most awful and momentous period I had hitherto experienced. The immediate hallooing of all hands on deck; to the pumps, plumb the well, cut away the masts, throw

I

the guns overboard. And amid all this activity, the dismal moans of some, the screams of the women.' The crew hurled everything moveable over the sides and the rising tide finally pulled the ship back to the sea's uneasy safety. Once back in Portsmouth, Manby reflected on his experience and concluded that the sea and he were not well suited to each other. Instead, he applied for a post as barrack master, a position which allowed him to keep his military honours while staying safely on dry land.

Manby did admittedly have good reason to be cautious. Two centuries ago, almost a third of all British seamen died pursuing their trade, being either killed by the punishment of life on board ship or sacrificed to storms and drownings. Nearly everything the modern mariner relies upon – competent maps, accurate instruments and adequate communication – was either unreliable or non-existent. The major sea lanes around Britain were crowded and collisions were frequent. What is now fixed and understood was then debatable, and navigation was more a matter of art than science. Sailors depended on experience or luck to avoid danger, and when they did run into trouble there was no kindly lifeboat service to deliver them. Until the mid-nineteenth century, it was made harder to assist victims than it was to collect the proceeds from wrecks. Previous legislation had defended the salvagers, not the mariners, and neither government nor shipowners devoted much attention to the consequences of nautical disaster. Most efforts were aimed at protecting cargo rather than ensuring that the crew returned intact with the goods. For several long centuries, lives lost at sea were regarded by much of Europe as so much natural wastage. Accounts still exist of sailors watching slack-handed from the gunwhales while one of their colleagues drowned. Once a person had fallen overboard, so the thinking went, he had been claimed by the sea, and it was not for mankind to challenge that claim. Such superstition was only an ideological response to an uncomfortable fact; the sea did kill people in great numbers, year after year. And, short of refusing to leave the safe shores of Britain, there was almost nothing that could be done about it.

Manby seized on the belief that something more must be done to prevent the deaths of shipwreck victims beached on the indifferent shores of Britain, if not for compassionate reasons, then certainly for civilised ones. The destruction of the gun-brig *Snipe* off the coast of Yarmouth in 1807 only confirmed his views. One hundred and forty-four lives were lost after the ship ran aground during a gale less than one hundred yards from the shore. Manby watched the ship beat itself to death on the rocks, and listened impotently to the cries of those still on board as they died. Over the next few months, he began experimenting with possible solutions. He concentrated his efforts on the idea of throwing a line from the shore to a distressed ship, using a rope fixed to the end of a canon ball. Several early versions failed spectacularly: the rope was either burned through by the gunpowder, or, in those rare instances when the ball and rope successfully reached their target, only managed to set what remained of the ship on fire. At the same time, he tinkered with the notion of an unsinkable boat. During a storm, small row-boats which were used to ferry survivors from the wreck to the shore almost invariably sank, either capsized by the seas or flooded by waves. Manby sealed several small wooden barrels with pitch and fixed them to the sides of a small, undecked boat, providing primitive but workable buoyancy chambers.

By the summer of 1807, his prototype mortar line was ready for testing. Until then, his colleagues and neighbours had watched Manby's eccentric experiments with derision. But once he produced something that threatened the wreckers, who took their livelihoods from the plunder of injured ships, he became a more serious danger. As the wreckers saw it, he was not only removing their prized source of income, he was also directly contradicting the will of God. God, they reasoned, had sent the storm that had wrecked the ship which they took as their reward. Any interference was therefore a form of devilish meddling. And so, helped by the knowledge that Manby could not swim, the wreckers tried to drown him. Several local sailors volunteered to help Manby demonstrate his boat and mortar line, and, when the boat was a good way from the shore, deliberately capsized it. Manby, just managing to keep afloat, was finally saved by his own efforts and two conscience-stricken spectators.

His determination was, if anything, increased. In 1808, the small brig *Elizabeth*, carrying a crew of seven sailors, became snared on the sandbanks near Yarmouth during a storm. Manby, seizing his opportunity, positioned himself and his mortar line on the shore and managed to fire the line successfully onto the ship. To do so required accurate calculation, since visibility was minimal and the mortar risked being damaged by both rain and salt water. Manby fired; the rope coiled outwards to sea. There was a pause, and then, slowly, the rope tautened. Someone, it seemed, was still alive and had taken up the line. Once he was confident that the link from ship to shore was secure, Manby sent out one of his ballasted row-boats. When it finally groped its way back to shore, seven sodden survivors crawled exhausted onto the beach. As they later testified, all of them would have almost certainly drowned without the help of the mortar line. Manby, all his heroic intentions vindicated, knelt on the sand and wept.

During the next forty years, it was later calculated, upwards of 1,000 people were rescued from disabled ships by the mortar line. But if Manby thought that his battles had ended there on the darkened beach, he was wrong. His inventions did not meet with the esteem he had hoped for. Despite his energetic lobbying in Parliament, the Admiralty remained suspicious, and the local sailors agitated loudly for his removal. Other, more credible competitors came forward, claiming either to have invented the line before Manby or to have improved it far beyond his original designs. Even his later attempts to establish a lifeboat service to rescue those victims of shipwreck who were stranded farther out to sea met with disapproval or silence. Manby was present at the inaugural meeting of the Royal Society for the Preservation of Life from Shipwreck in 1824, at which the first decision was to award five gold medals: one to the King, one to the Duke of York, one to the Archbishop of Canterbury, and one each to Manby and Sir William Hillary. Manby took the medal, but not the credit. Hillary, a boastful but enterprising Yorkshireman whose career bore a strong resemblance to Manby's, is now generally honoured as the founding father of the Royal National Lifeboat Institution. Admittedly, Manby did not always help his own cause. He could be pompous and grasping, and was hopelessly vain. His journals and letters are littered with moments when he 'considered his fortune made' by the platitudes of a minister or the flattery of a courtier. His efforts often had more to do with his own self-advancement than they did with more generous motives. But for all his deluded grandeur, Manby did achieve great things. After the invention of the mortar line, he spent the rest of his life searching for the recognition he felt he deserved. Just before his death aged eighty-nine, he decided to build a monument commemorating both himself and the mortar line. When completed, he offered it to the town of Yarmouth, which had already built and paid for a statue of Nelson in the town centre. Manby's memorial was rejected; he was left with nowhere to put it but his own front garden.

From Berwick to the Solway, including the sea lochs of the west, the Scottish coast runs for 4,467 nautical miles. By the late eighteenth century, that coast had gained an ominous reputation. Most mariners stayed well clear, and those who sailed there often chose to continue travelling all night instead of looking for landfall after dark. Not only is Scotland girdled by two opposing seas – the North Sea and the Atlantic – but her ragged island archipelago provides a major obstacle course for the best of sailors, even now. In the past, the sea was considered so dangerous in winter that one early Act had forbidden ships stocked with essential goods from leaving Scottish ports between the end of October and the beginning of February. The reasoning was obvious. As well as the sandbanks and treachery of the English coast, Scotland is moated by an awkward brew of conflicting tides and currents. The North Sea, which in the eighteenth century gave the only access to the Continent, Scandinavia and Russia, is a dark place of streams and sudden climatic switches. In the Pentland Firth, where the North Sea meets the Atlantic, sailors face riptides, cross winds and breakwaters on the water, and sandbanks, skerries and reefs underneath. Often, the competing tides set up currents that run at ten knots or more, each of which is troublesome enough to have earned its own title: Duncansby Bore, the Merry Men of Mey or the Swilkie. The Pentland Firth itself is still known simply as 'Hell's Mouth'. The names that still speckle marine maps represent more than just picturesque history. The Black Deeps and Blackstone Banks took their percentage of dead, year after year. Farther south, the Solway Firth is riddled with quicksands and the Firths of Tay and Forth are notorious for rocks. In the 1870s, Thomas Stevenson, a man not given to hyperbole, calculated the pressure of the breakers on the coast at up to 6,000 pounds per square foot.

Even when captains were confident they were clear of the cross-currents near the isles, there were reefs and rocks – Sule Skerry off Cape Wrath, the Torran Rocks off the coast of Iona and Skerryvore beyond Tiree. The entrance to the Forth is guarded by three separate obstacles – the Isle of May, the Bass Rock and Inchkeith – all of which stick directly in the path of shipping on its way to the port of Leith. A little farther northeast, there is the infamous Bell Rock, submerged at high tide and a confusion of serrated rocks at low tide. Round on the west coast, the entrance to the Clyde is calmer but prone to shallows and awkward weather patterns thrown off by the surrounding islands. Where the placid Gulf Stream soaks into the Atlantic, the area is notorious for races, currents and 'standing waves' (water flowing over submerged objects which gives the impression of immobility), while the whirlpool of Corrievreckan between Jura and Scarba is considered the most dangerous tide in Europe. Even on a clear day, crossing Corrievreckan involves careful calculation. Boats making the passage cling nervously to the shoreline only to find themselves speeding through at fifteen knots or more. In a dirty sea, the gully resembles Scylla and Charybdis, sucking boats down into the eddy or spitting them out to the shores on either side. Sailors who know the area well enough to avoid the centre frequently wreck themselves on the nearby coasts trying to escape.

Until the mid-nineteenth century, navigation remained a ramshackle skill. Sailors within sight of the shore depended on being able to recognise the coastline. A church roof, a solitary copice or a coastal hamlet were reference points as dependable as any more thorough knowledge of the sea's geography. In England, Elizabeth I had made altering or dismantling the most significant coastal landmarks a criminal offence. The law had little effect. After dark, without lights, landmarks were of only the most limited help. Many accounts of shipwrecks from the time note laconically that the pilot had mistaken one bay or harbour for another, and ended up paying for it. Even when the first primitive fires were lit on headlands to mark the way, they could easily be confused with stubble fires or temporary beacons. Nor was it possible to rely on written evidence. Even now, with the benefits of sonar and satellite surveying, there is no such thing as a definitive chart. Some parts of the Scottish coastline have not been surveyed for 150 years or more; others could be surveyed till the end of time and still not keep pace with the shiftiness of the sea. In the late eighteenth century, charts, maps and pilot books were drawn up by trial and painful error, and they were as often produced by merchants or traders as by any regularised state system. It was not until 1750 that Murdoch Mackenzie published a sea chart of Orkney and Lewis based on a rigid triangulation framework. Farther south, the situation remained poor for several more decades. In 1788, Murdo Downie, the Master of HM *Champion* at Leith, was complaining to the government that he could find 'no chart published of the East Coast of Scotland that could in any degree be relied upon'. The old cherub-covered maps, with their foreshortened coasts and squint-eyed headlands, might look endearing now, but for several long centuries they were the only detailed information on the British coast available. When Daniel Defoe made his tour of Scotland in the 1720s, he discovered that the Forth and Clyde did not, in fact, run into one another, as his map suggested. 'When I came more critically to survey the ground,' he complained, 'I found the map-makers greatly mistaken, and that they had not only given the situation and courses of the rivers wrong, but the distances also.' Farther north, in 'that mountainous, barren and frightful country', the Highlands, things were even worse. 'Our geographers seem to be almost as much at a loss in the description of this north part of Scotland, as the Romans were to conquer it,' Defoe noted disconsolately, 'and they are obliged to fill it up with hills and mountains, as they do the inner parts of Africa with lions and elephants, for want of knowing what else to place there.'

Matters improved erratically, if at all, in the nineteenth century. In 1837, an indignant committee of Edinburgh societies complained to the Treasury that even the best maps and charts of Scotland were so inaccurate that 'in some charts, the large island of Arran is laid down as six miles from Bute, in others as nine miles, and in a third as 12 miles distant from the island. Pladda Island light in charts is placed at 16° north of Ailsa Craig, where its true distance is only 10°20'. These last are serious errors at the entrance of so important a river as the Clyde.' Many of the roughest hazards remained unmarked, those that were noted were often wrongly placed, and the pilot-age rules, with their 'Fifty fathoms black ooze and black fishey stones among', could often be more poetic than practical. Areas given as 'safe anchoring' were revealed to be notorious ship-wreck spots; ports and harbours were awkward to approach and littered with the bones of old ships.

Not that adversity deterred Scotland's swelling population from turning away from the land and onto the sea. Like the rest of Britain, Scotland needed it, fed off it, took employment from it and profited by it. From the sixteenth century onwards, the nautical traffic around Britain increased steadily, while the numbers of shipwrecks and groundings rose in tandem. Aside from farming and manufactures, the sea provided one of the principal sources of employment for a large swathe of the popu-lation until well into the nineteenth century. Directly, it pro-vided subsistence, fish and trade; indirectly, it provided strength, funds and political muscle. By the beginning of the eighteenth century, Britain was also spending a significant amount of her time and money waging wars across it. The navy grew threefold, and with it grew the pirates, privateers and press gangs of legend. Though the end of the Napoleonic wars meant the dwindling or abolition of all three, for the moment they remained a con-stant threat. The escalation of trade meant the escalation in war to protect that trade. In the century between 1650 and 1750, England was engaged in six major European wars. Old routes were travelled more frequently; new ones were marked out. By the 1750s, Scotland and England had separately built up a regu-lar trade in subsistence goods – corn, coal, livestock – with France, Scandinavia and the Baltic. The French, meanwhile, were involved in so many wars at the time that they were forced to scale down their navy and resort to privateers instead, many of whom spent their time raiding the British coast. At the same time, Scotland in particular depended on the Scandinavian countries as trading partners and maritime allies. The traffic between the two places, always constant, escalated with the growth of industry and the spread of free trade. From the Clyde ports there was the journey to the New World, which by the 1750s was providing a useful new source of tobacco, sugar, manufactured goods and slaves. To the north and south, there were the whaling grounds and beyond the Continent there were the exotic dangers of a new empire. In each direction, there were prizes to be taken and claims to be staked.

Scotland contributed her own heavy percentages to the traffic in other ways. Union with England had brought benefits, albeit slowly. Immediately after 1707, the changes were mainly internal: cattle sales to English markets, corn to English mills or men to English employment. But after the 1745 Jacobite Rebellion, Scotland's trade with Europe accelerated and the age of the Great Improvers began. Landlords in the Highlands cleared the straths for sheep, packed off the protesting population to stony coastal settlements, taught them how to fish and left them to make a life for themselves. Some of the settlements died quietly, others took root and became export centres for wool, flax or fish. The subjugated regiments joined English wars and British battles or went south to the new shipbuilding yards in Glasgow or Greenock. Several thousand Highlanders left for the New World on leaky ships, some of which did not last the journey. Edinburgh functioned as both Continental trading post and garrison town, providing goods for export and men for war. When the press gangs sought fodder for their frigates, they looked first to the Scottish capital. For whatever cause, the population of Scotland was on the move in a way that it had not been before, and much of that movement was by sea.

And as the sea cluttered up with shipping, so it accumulated shipwrecks. In the 1790s, an average of 550 ships were wrecked every year on British shores; by the 1830s, the numbers had risen to well over two a day. The vast increase in nautical traffic around Europe had not yet been matched by any improvement in safety. There was no regulated distress code and only the most clumsy and primitive of aids: heavy leather lifejackets or inadequate row-boats. By 1800, Lloyds of London estimated that one ship was lost or wrecked every day around Britain; between 1854 and 1879, almost 50,000 wrecks were registered. The figure is probably ludicrously low. Many wrecks never reached the attention of the local Admiralty officer, either through difficulties in communication or, more likely, through deliberate concealment. Both the navy and the merchant ship-owners learned through bitter experience to expect a certain percentage of their ships to sink every year they sailed. With the mortality rate so high and conditions so bad, the sailors themselves could only cultivate a brutal fatalism about their work. They lived in a twilit world, with their own jargon and codes; most did not expect to live beyond the age of forty. They regarded the government with suspicion, the law with indifference, and their landlubber compatriots with derision. They were accustomed to shipwreck or injury, they accepted that the sea was unsafe, and they remained suspicious of men who promised salvation.

Given such an ominous background, it was evident that changes would have to be made. By the 1780s, the swell of public agitation had become too strong for the government to ignore any longer. But it is notable that the pressure for light-houses did not emerge from the sailors most at risk or indeed the organisations best equipped to provide lights. The pressure came from the shipowners and the naval captains, both of whom were keen to minimise the risks in sending precious cargo to sea. Their crews, the men who did the dying, seemed either so pessimistic about their chances for survival or so sceptical of innovation that it took several decades to convince them of the need for lights. It was the captains, the money men and the merchants who agitated most fiercely for action. Finance, as usual, took precedence over compassion. But out of such necessity came something more exceptional than the usual desultory efforts to mark the places between the sea and the land. Though the drive to build the Scottish lights was commonplace enough, the men who came to fill that role were made of rarer stuff.

TWO

Northern Lights

For a short while, it seemed as if Robert Louis Stevenson might fulfil his parents' ambitions. For nearly twenty years, he had been a worry to them. Now, when it came to settling down, he alarmed them even more. First there had been the sickliness, then the lack of schooling, then the whispers of midnight societies and shady liaisons. Worst of all was Louis's terrible wandering mind. He seemed not to stick at anything, and spent most of his time aimlessly pacing the streets of Edinburgh or dabbling in books. His mother pleaded, and his father, Tom, grew neurotic with worry. Louis, guilty and cross, avoided home. Finally, in the spring of 1868, Tom persuaded his son that it was time he applied himself properly to the family business. Louis was to be enrolled at the University of Edinburgh to study civil engineering, and would spend his summer vacations serving his apprenticeship at his father's projects around the country. First he was to go to the harbourworks at Anstruther and Wick, then on the lighthouse steamer's journey round Orkney and Shetland and finally he would supervise the Dhu Heartach lighthouse works on the Isle of Earraid. Whether he liked it or not, he would follow in his father's footsteps, just as Tom had followed in Robert's. Louis capitulated and for a while his parents stopped fretting.

It did not last long. For three long summers around the northern shores of Scotland, Louis tried to bend his mind to the disciplines of engineering. Tom received erratic reports of progress, the news of an underwater trip in a diving bell, and occasional muffled cursings at the intransigence of the weather or the incompetence of the workmen. Louis experimented with waves, fussed over the slowness of his drawing and tried without conviction to improve his mathematics. 'My daily life,' he told his cousin Bob gloomily, 'is one repression from beginning to end.' While Tom continued to receive news of the slow progress of building at Dhu Heartach, Louis spent the rest of his leisure time wistfully discussing metrical narratives and small beer in letters to friends. In the spring of 1871, back in Edinburgh, Louis presented a paper, 'On a New Form of Intermittent Light for Lighthouses', at the Royal Scottish Society of Arts. The essay showed the accumulated knowledge of three obedient years following the Stevenson grail: it was workmanlike, efficient, and showed no spark of initiative whatsoever.

Tom was among the audience and watched Louis being awarded the Society's silver medal. For him, it was a proud and vindicating moment; Louis, it seemed, had finally submitted to good sense. A week later, the two walked out to Cramond. 'On being tightly cross-questioned,' wrote Louis later, 'I owned that I cared for nothing but literature. My father said that was no profession.' Angry and desperate, Tom suggested something else instead, 'and so, at the age of 21, I began to study law.' It was small consolation for both of them since Louis was no more interested in advocacy than he was in engineering. Tom was left to blame his son's fall from grace on a surfeit of imagination and too many books. Later, the two fell out even more dramatically over Louis's agnosticism, but even then never completely separated. For years, Tom continued to send his son corrective notes on his fiction; a little more Scripture here, a little sermonising there. Sensibly, Louis ignored him. But it was a measure of Tom's affection that he abandoned his engineering ambitions for Louis with so little resistance. As Maggie Stevenson, Louis's mother, later noted, 'it was a cutting-short of his own life, as he had looked forward to its being continued in his son's career.'

Louis, it seemed, had been quick to recognise both the benefits and drawbacks of his family's profession. As he wrote in *The Education of an Engineer*,

> It takes a man into the open air; it keeps him hanging about harbour sides, which is the richest sort of idling; it carries him to wild islands, it gives him a taste of the genial dangers of the sea; it supplies him with dexterities to exercise; it makes demands upon his ingenuity; it will go far to cure him of any taste (if ever he had one) for the miserable life of cities. And when it has done so it carries him back and shuts him in an office! From the roaring skerry and the wet thwart of the tossing boat, he passes to the stool and desk; and with a memory full of ships, and seas, and perilous headlands, and the shining pharos, he must apply his long-sighted eyes to the petty niceties of drawing, or measure his inaccurate mind with several pages of consecutive figures. He is a wise youth, to be sure, who can balance one part of genuine life against two parts of drudgery between four walls and for the sake of the one, manfully accept the other.

Later, still smitten with guilt over his exile from Scotland and his family, he wrote a revealing poem.

> Say not of me that weakly I declined
> The labours of my sires, and fled the sea,
> The towers we built and the lamps we lit,
> To play at home with paper like a child.
> But rather say: *In the afternoon of time*
> *A strenuous family dusted from its hands*
> *The sand of granite, and beholding far*
> *Along the sounding coast its pyramids*
> *And tall memorials catch the dying sun,*
> *Smiled well content, and to this childish task*
> *Around the fire addressed its evening hours.*

In practice, the idea of Louis as an engineer was absurd; he was far too physically frail to have lived the working life of his father and grandfather. But he remained haunted by the notion that his writer's life was somehow less noble or worthy than the rest of his family's more practical achievements.

One of Louis's many attempts to redress the balance was in an unfinished Stevenson biography, *Records of a Family of Engineers*. The early Stevensons, he discovered, had supplied nothing but generation upon generation of tenant farmers, with the exception of John, a seventeenth-century ancestor and 'eminently pious man' who seemed determined on Protestant martyrdom. John spent 'four months in the coldest season of the year in a haystack in my father's garden' and sleeping in Carrick fields under a blanket of snow. Though he did contract scrofula, he was spared persecution, to his apparent disappointment, in the religious purges of the 1680s. With the exception of John, however, Louis's genealogy was one of stolid mediocrity. 'On the whole,' he wrote, 'the Stevensons may be described as decent reputable folk, following honest trades – millers, maltsters and doctors, playing the character parts in the Waverley Novels with propriety, if without distinction, and to an orphan looking about him in the world for a potential ancestry, offering a plain and quite unadorned refuge, equally free from shame and glory.' In the absence of glamorous fact, Louis felt himself forced to resort to speculation. He considered the possibility of a Scandinavian connection, evidence of a French alliance and, more imaginatively, the link with a Jacobite past. By the time Louis had completed his history, the family had acquired a smattering of Highland credibility and a link with that most glamorous of cattle-rustlers, Rob Roy MacGregor. Later biographers noted crushingly that none of this wishful thinking was true. The Stevensons were descended from quiet Lowland Whigs, none of whom ever had a dangerous political thought in their lives.

Louis's real interest in the Stevensons began with the birth of his paternal grandfather, Robert Stevenson. Robert's father,

Alan, was a Glasgow maltster who married the daughter of a builder, Jean Lillie, in 1771. On 8 June 1772, their only son was born. Alan was still a young man, barely twenty, and with his brother Hugh had become involved in the Glasgow trade with the West Indies. When Robert was two, his father and uncle sailed south to look after their business interests, leaving Jean and Robert behind in Glasgow. Once in the Caribbean, the Stevensons found themselves the victims of a swindle. One dark night, two local merchants – possibly business competitors – arrived at their house on St Kitts, and robbed them of the contents. As soon as they heard of the burglary, Hugh set sail in pursuit of the robbers, while Alan remained behind to deal with the business. 'What with anxiety of mind,' Robert later recorded, 'being such very young men – and exposure to night dews of that climate, the two brothers were seized with fever and died in 1774, my uncle at Tobago on 16 April and my father at St Christopher on 26 May.'

'Night dews' was then the catch-all diagnosis for any tropical disease that British science had not yet explained or cured. Malaria, cholera and tuberculosis were rife, as was sleeping sickness and influenza. Whatever the cause, the consequences of Alan's death were, for Jean Lillie, terrible. While still young, she was left a widow with a small child, short of money and dependent on her mother for subsistence. But despite her sudden poverty, she showed a fierce loyalty to her only child. If she could not improve her own circumstances, she reasoned, at least she could improve Robert's. Her father had sent her to an Edinburgh boarding school and Jean clearly felt the benefits of an Edinburgh education, so taking her six-year-old son, she moved the forty miles eastwards to the capital. When the time came, she tried to enrol Robert at the High School (where Walter Scott and Henry Cockburn were being educated), but found she could not afford the fees. So she enrolled him at an endowed school and kept aside a little money to pay for extra tuition in the classics. Robert's upbringing therefore became a stern apprenticeship in scrimping interspersed with plenty of Latin and God. In the mercantile freedom of the 1780s, Robert was taught the essential details: to put his faith in hard work, merito-cracy and the middle-class world. For a while, his mother hoped that Robert would make a minister of the Church of Scotland. Fortunately, his lack of Greek and hopelessness at Latin put paid to the idea.

Once established in Edinburgh, Jean Lillie began attending church in the New Town. Among the congregation was another family, the Smiths. In early middle age, Thomas Smith was a stout man, tall, plain and pragmatic. He had come originally from Broughty Ferry – then a briny little suburb of Dundee – and, like Jean, had been forced to learn self-reliance early. When Thomas was still young, his father was drowned in Dundee harbour. His mother, left with a small child, brought him up herself as best she could, gave him a good and pious education and insisted that whatever trade he took, it should at least be safely on land. Thomas took his mother's instructions to heart and found an apprenticeship with a Dundee metal worker where he spent the next five years learning ironmongery before moving down to Edinburgh. After a few years on the staff of a metal-working company, he established his own business in Bristo Street as an ironsmith, providing grates, lamps and intricate trinkets for the New Town. The business throve and Thomas prospered. He was the creature of a most particular time, a high Tory and a businessman of talent and ambition. Louis considered him 'ardent, passionate, practical, designed for affairs and prospering in them far beyond the average'. His first wife was a farmer's daughter who bore him five children. Despite Edinburgh's reputation for medicine, surviving the rigours of childhood in the eighteenth century was still a matter of good fortune. Thomas's first children were not lucky. Three of the five babies died; only Jane and James survived. His wife too finally succumbed to whooping cough, and Thomas was left a widower with two small children. He was married again, in

1787, to the daughter of a Stirling builder who bore him one daughter, Mary Anne, and then promptly expired of consumption. Thomas, now well accustomed to mortality, started looking for another wife.

Jean Lillie, with her small, well-disciplined son and her belief in similar ideals, was a willing match. In June 1787, the two were married. It was a pragmatic partnership, based as much on the benefits of uniting two incomplete families as on affection or companionship. It was also a marriage of equals. Jean was a strong-minded soul, who earned the devoted respect of her new husband in return for security and a stable upbringing for his children. The arrangement also served Robert well. By now, it was evident that he had infinitely more talent, and greater patience, for the practicalities of mechanics than he had for Latin. When Thomas took him on a guided tour of his ironworks, Robert was beguiled by the blend of craftsmanship and usefulness. By 1790, Thomas had taken Robert on as an apprentice, and the boy's efforts at Greek, French and theology were forgotten.

Thomas's main business at the time was in lamp-making and in designing street lighting for the New Town. The lamps at the time used oil, which silted up quickly with grime and gave out only a weak and smoggy light. Thomas, experimenting with methods of improving the standard design, began devising a system of reflectors placed behind the light to strengthen and focus the beam. The idea, he knew, had been used successfully elsewhere in Europe but had not yet been introduced to Scotland. At first, he made the reflectors out of concave circles of copper, the size and shape of scooped-out melons and polished to a high sheen. Later, he varied the design, welding small slivers of mirror to the back of a concave lead circle. Seen now, his reflectors look like an inside-out antique mirrored ball, but in 1780s Edinburgh, they were revolutionary. All light sources were measured in units of candela (or candlepower), one medium-sized tallow candle producing roughly two units of power. Thomas's designs quadrupled the strength of the light and produced something closer to a concentrated beam than a lamp on its own could ever achieve.

Thomas fitted several of his parabolic reflector lights around the New Town and then, mindful of the need for more business, contacted the Trustees for Manufactures in Scotland. His reflectors, it seemed, had practical applications beyond mere street lighting; would they, enquired Thomas, be useful for lighthouses? As Thomas explained it to them,

> Lamps being Inclosed are preserved from the Violence of the wind and weather but Coal lights cannot be inclosed ... Lamp light of itself has a more pure and bright flame than Coal light and when Conjoined with a reflector of proper power transcends it ixceedingly and is seen at a much greater distance ... Lamps take less attendance ... Lamp lights with reflectors can be distinguished from every Other light in Such a manner as to make it Impossible to mistake them for a light on shore or on board any Other Ship ... Coal lights are not capable of this Improvement.

He had, he declared, already prepared a sample reflector and was happy to demonstrate it to 'any gentleman concerned'. The Trustees inspected his work, and agreed that Thomas's designs would indeed be useful. Beguiled by his enthusiasm, they sent him south to gain experience from an English lighthouse builder. Once returned, he was made first engineer to the Northern Lighthouse Trust.

The title might have been imposing; the organisation itself was not. The Trust (now the Northern Lighthouse Board, or NLB) had been established after complaints about the state of the Scottish coast had reached boiling point. Several evil-tempered storms during the winter of 1782 had crippled both the naval and the merchant fleets, both of whom urgently petitioned Parliament to remedy the existing situation. Parliament set up a committee which recommended the construction of at least four lighthouses, scattered at vital points around the Scottish coast. In 1786, the bill was passed and the NLB was born. The Act for Erecting Certain Lighthouses in the Northern Parts of Great Britain provided for a management committee and a few officials to collect revenue, stipulated the sites of each light and then left the Board to its own devices. With its leaden cargo of sheriffs, judges and public goodbodies, the committee had only the most feeble knowledge of the sea and none at all about lighthouse construction. The Lighthouse Commissioners, as they were known, had been selected with the intention of providing political and financial canniness to the Board; as public officials, they were accustomed to question the cost of everything and trust the value of nothing. Their appointment was also partly political. Since the lighthouses would be built within the sheriffs' fiefdoms, it was easier to give each one a place on the Board than to woo them anew every time a light was to be built on their land.

From the beginning, therefore, there was a strict division of roles. The Commissioners figured and the Engineer built. The Commissioners respected the experience and integrity of the Engineer; the Engineer lived within the Commissioner's restrictions. 'I beg leave to acquaint you that I am willing to do every thing in my power to bring to perfection the plan proposed,' wrote Thomas in September, 'and to superintend the erection of the lights, teach the people how to manage them and to do every thing necessary to put and keep them in a proper state ... It is impossible at present to form any judgement of the trouble attending this business as I hope it will turn out a benefit to the Country.' His first job, they announced, would be to construct four new lights: one at Kinnaird Head just beyond Fraserburgh, one at Mull of Kintyre overlooking the Firth of Clyde, one at Eilean Glas off the edge of Harris, and one on North Ronaldsay, a small island above the Orkney mainland. He could build them, staff them and light them in any way that he wished, they promised, as long as it wasn't too expensive. The Commissioners, meanwhile, stayed in Edinburgh, counting the costs and squabbling politely with London.

Thomas could have been forgiven for wondering what exactly he had got himself into. For a start, he had no architectural experience, let alone the kind of experience necessary to build a sea-tower exposed to the hardest conditions wind and wave could hurl at it. Lighthouse construction was, to say the least, a specialist subject in the 1780s, although the idea of a lighted tower for mariners' guidance had existed in some form or another since the Pharos of Alexandria was built by the Egyptians in 300 BC. The Pharos was an immense and ornate tower 450 feet high, topped with an open fire, and considered so splendid that it was usually listed as one of the Seven Wonders of the World. Later attempts were less glamorous. Since there was no state control of lighthouses in Britain until well into the eighteenth century, their design was left to the individuals who built them. Far from being the trim marine spires of popular image, the English lights developed endless exotic variants. Most were coastal towers with large braziers of coal fixed to their roof. Some were church steeples loosely adapted for the purpose. Disused castles and priories were occasionally put to use, and in Ireland, there were several lights built in stone-vaulted cottages. Even those built specifically as lighthouses did not follow any definite pattern. Some looked like stumpy medieval rockets, some like upright coffins, others not much different from the average cow byre. One or two followed the design of fortified keeps, sturdy enough for the hardest gale. Others were no more than an iron basket filled with coal and suspended on a pulley. A number of owners built their lights in wood. Unsurprisingly, not many examples of these survive.

Until the 1780s, the only permanent light in Scotland was on the Isle of May which had been alternately saving and exasperating mariners for a century or more. The lighthouse had been built in 1636 on a small, low-lying islet at the entrance to the Firth of Forth. The mouth of the river was filled with snags for unwary shipping; rocks, sandbanks and awkward reefs on the surface and a graveyard of dead ships underneath. The islet was the first and the largest of these rocks and had gained a vicious reputation for shipwreck and destruction. In 1635, the Scottish Privy Council had given the task of constructing a light to three of Charles I's favoured Scottish courtiers, who designed it, built it and maintained it at their own expense and then charged local shipping for its use.

Even by the make-do standards of the age, conditions for the lone keeper were unusually grim. The isle, a mile long by a third of a mile wide, was barren except for a little pasturage and a low, squat tower like a medieval keep with a brazier on the roof. The owners hired a local man, George Anderson, to tend and supervise the fire, and arranged for a boatman to appear every few days to drop a new delivery of coal into the shallow waters by the island's rocky shore. Anderson would pick the coal out of the waters, haul it along to the tower on his back and winch it up in a bucket to the roof. For this, he was given a salary of £7 a year, 30 bushels of meal for his family and all the fishing rights he could want. Since he therefore spent most of his time away from the island looking for fish, the fire stayed untended and usually went out at the crucial moment. It was several years before someone took pity on the poor man and offered him the assistance of a second keeper and a horse.

Even when the fire was maintained, one light was hardly satisfactory for all Scotland. As the Isle of May proved, coal lights were inefficient, gobbled fuel and expired just when they were most needed, in gales or heavy rain. They required constant supervision, were usually clogged by smoke and could easily be mistaken for fires inland. Thomas was evidently going to have to start from scratch, devising new buildings, new fuels and new solutions if he was to succeed in improving the current situation. But if lighthouses came without templates, so too did their architects. There was, as yet, no such thing as an archetypal engineer, let alone a civil or marine specialist. The qualifications and bureaucracy of the modern profession did not exist 200 years ago. When Samuel Johnson published his famous Dictionary in 1755, he described an engineer as 'an officer in the army or fortified place, whose business is to inspect attacks, defences, works'. As Louis later pointed out, 'the engineer of today is confronted with a library of acquired results; tables, and formulae to the value of folios-full have been calculated and recorded; and the student finds everywhere in front of him the footprints of the pioneers. In the eighteenth century the field was unexplored; the engineer must read with his own eyes the face of nature; he arose a volunteer, from the workshop or the mill, to undertake works which were at once inventions and adventures.' If he was to design, build, supervise, and maintain each of the NLB's new lights, Thomas therefore needed to become an inventor in his own right. Much of his work was without precedent, and where tested methods did not already exist, Thomas had to improvise as best he could.

Trudging around the rim of Scotland, he soon realised that his new role entailed far more than merely fitting reflectors. Initially, he used the English lighthouses as his template, but was forced almost immediately to adapt to Scotland's particular rigours. Most English lights of the time were built safely inland out of dependable local stone. Any building on the stormy coast of Scotland needed to withstand all that the elements could hurl at it; a lighthouse, with its flimsy glass and curlicues of ironwork, required a particular kind of strength. The first four lights were, according to Robert Stevenson, 'on the smallest, plainest and most simple plan that could be devised, and with such materials as could be easily transported and most speedily erected'. All were built of unembellished stone, with walls thick enough to resist the strongest assaults of water or wind, and with plain lanterns bound with a tight corsetry of metal stanchions. The light at Kinnaird Head was adapted from an old fortified tower, and did just as well withstanding the elements as it had done withstanding invasion.

Despite the economy of the designs, Thomas soon discovered that his workload doubled. His assistants were untrained and his experience was suited more to the refinements of New Town ironwork than it was to designing weather-beaten coastal buildings. Two of the four planned lighthouses were on remote islands, which meant long, dangerous sea journeys and difficulties with supervision. Even on the Mull of Kintyre, which was at least on the mainland, materials could not be landed on the coast and therefore had to be carried over twelve miles of stark moorland to the site. Every slate, every block of stone, every fragment of lens and gallon of fuel, had to be taken on horseback and even then the journey was considered so difficult that only one trip could be made each day. Moreover, many of the materials for the lighthouses were new and untried. It was difficult for instance to manufacture glass large and strong enough for the storm-stressed lanterns, since reinforced glass had not yet been invented and metal supports would only have obscured the light.

Many of Thomas's experiments worked well in his Edinburgh workshop, but when transported out to the edges of Scotland were found to be impractical or unusable. Where possible, he used local building materials, with the trusted Scottish combination of granite, slate and wood as the basis of the buildings. The delicate mirrored reflectors, however, had to be transported from his Bristo Street workshop by sea to the site, and were sometimes found to be ill-fitted for their purpose. The numbers of reflectors in each light had to be varied, and each one brought its own difficulties of transportation and installation. For a long while the task of constructing the lights seemed so impossible that Thomas had considerable difficulty persuading the incredulous local builders to work for him.

Appointing the first lighthouse keepers also presented unexpected difficulties. In the early years, Thomas looked mainly for retired shipmasters and mariners, either hired locally or brought to him by word of mouth. They too had to be pioneers of a kind; much of their role could only be resolved through experience, and a precise definition of their duties only emerged over time. But they were also responsible for a great deal of delicate and expensive equipment, and Thomas left little to chance. James Park, the first keeper at Kinnaird Head, was instructed to

> clean the Reflectors and the panes in the windows every day the proper manner of cleaning the Reflectors is to take off any Oil or Smoke that may be found upon them with soft tow and then rub them with a soft linnen rag and Spanish white or finely pounded Chalk till they are perfectly bright this must be strictly attended to or else a great part of the effect of the lights will be lost ... You will light the lamps half an hour after Sun-seting and keep them burning till half an hour before Sunrising every day for which purpose you must attend them every two or three hours throughout the night to help any of the lights that may be turning dim but you must take care not to stand before the lights any longer than is necessary of that purpose and you are to observe that in stormy weather you must not leave the light room the whole night.

In return, Park was given a shilling per night, free lodging, and pasturage for a cow. He remained contentedly in the job for a decade, before retiring aged almost eighty.

There were also some staffing difficulties Thomas could not have foreseen. At North Ronaldsay, the keeper took good relations with his neighbours too far, and had started his own local black market in lighthouse fuel. 'The keeper,' Smith wrote indignantly in his report to the NLB, 'has acted the most dishonest and infamous part that can be imagined. He has by his own confessions before a number of witnesses sold the oil sent him in very great quantities throughout North Ronaldsay and the neighbouring island of Sanday, so that his conduct is notorious in the whole country.' Smith rapidly discovered that, despite all his efforts, the keepers themselves kept introducing an unwelcome dose of chaos.

Perhaps most striking of all was his pained discovery that not everyone wanted or encouraged the lighthouses. Thomas and his Stevenson successors found that they did not merely have to compete with primitive materials and impossible geography, they also found themselves at war with inertia, hostility, superstition and disbelief. They had to do battle with landowners and government to get the lights built, and they found themselves challenging the prejudices of those whom those lights were supposed to save. Many people did not believe that lighthouses would work; many believed they were too expensive, many saw them as a form of religious defiance. Many people simply did not see the need for them. During the original inquiry into the need for a light on the Isle of May back in 1635, all the predictable excuses had arisen: the light would be too weak to be seen, the shipowners would be financially broken by the charges, there was no need for a light, the rock itself was not dangerous.

There were also more imaginative protestations. John Cowtrey, a skipper from Largs, complained that a light on the Isle of May would only guide ships to destruction on the nearby Inchcape and Carr rocks. George Scot, another skipper from Dysart, complained that since the light would not be visible in a snowstorm, there was no point in having one. Richard Ross, a merchant in Bruntisland, thought that boats would always be wrecked on the isle, and that a light would make no difference. James Lochoir, a skipper from Kinghorn, believed that ships which ran aground on the isle were stupid, and no light on earth would save them from their own imbecility. Exactly the same crop of complaints arose with subsequent lights; even when the benefits were there for all to see, there were plenty of souls who resolutely refused to acknowledge their usefulness.

Down in England, the protests were even more elaborate. Curiously, many of those complaints came from the lighthouse service itself. Like the NLB, Trinity House was impoverished for much of its early existence, since the Crown had originally given it the authority to build lights, but not to charge for them. It therefore spent much of its life finding ingenious new ways to wriggle out of its duties. During the seventeenth-century debate over the construction of the lights at North and South Foreland near Dover, Trinity House objected on the grounds that lights would only alert foreign ships to the British coast. They complained bitterly of 'such costly follies as lighthouses ... The Goodwins [the notorious Goodwin Sands] are no more dangerous now than time out of mind they were, and lighthouses would never lull tempests, the real cause of shipwreck.' And, as they concluded with a final, divine flourish of illogic, 'If lighthouses had been of any service at the Forelands the Trinity House as guardians of the interests of the shipping would have put them there.'

The most serious threat to the lighthouses, and one which was to bother both Thomas and Robert for far longer, were the wreckers. They, unlike the shipowners, the skippers or even Trinity House, had a vested interest in ensuring that ships were destroyed. Many coastal villages staked their livelihoods on the exotic plunder to be found in dead and dying ships; the wreckers saw their lootings as a perk of nautical life and bitterly resented any attempts to interfere. As Thomas discovered, the wreckers were furious at the prospect of a safer sea.

The increase in shipping, and the consequent increase in shipwrecks, meant that by the late eighteenth century, they were thriving on salvage and theft. Wrecks were so frequent that many coastal populations had come to regard the cargo as their right. Sailors who survived gales and destruction were often murdered by locals within sight of the shore, and certain areas of the country became notorious for wreckers'

exploits. In the south, it was claimed by one early historian
that

> If a wreck happened to occur in Cornwall while Divine
> Service was being held, notice of it was given out from
> the pulpit by the parson. It is said of the wreckers, I
> know not with what truth, that the strongest among
> them would swim out through the breakers and drown
> the exhausted survivors by thrusting them under water
> as the poor wretches struggled, with failing strength, to
> reach the shore. There were even pious fanatics who
> went so far as to admonish the people that it was sinful
> to succour a vessel in distress upon the Sabbath; that it
> was, in fact, sinful to save life. On the other hand, refusal
> to do so was a proof of true religiousness since it showed
> that they realised it was God's will that the ship should
> sink and the crew perish.

Some shipwreck survivors would be saved by a stray compassionate soul; more often they were regarded as a dangerous
inconvenience and left to die. The wreckers worked furtively,
away from the censorship of officialdom, and did what they
could to prevent the local customs officers from becoming too
curious. In many cases, those officials were either easily corrupt-
ible or practised at turning a blind eye. Their leniency was
perhaps understandable, since in many areas of Britain the population desperately needed the sea's harvest. The Hebrides in
particular included islands naked of a single tree, and their
islanders were forced to import even the most basic materials for
life. They thus relied heavily on driftwood and wreck to build
their houses, make their boats, warm their families and cultivate
their food, and they regarded Thomas and his mirror-lamps as a
mortal threat not just to their livelihood but to their lives.

More awkwardly, the wreckers could, with some justification,
claim salvage from a ship as a legal right. Until 1852 and the
Customs Consolidation Act which appointed official Receivers

of Wreck, the law remained confused and uneasy, and little could be done to prevent thefts. Previously, all wrecks in British waters came under the jurisdiction of the Lord Admiral whose role was to take 'cognisance of the death of man, and mayhem done in great ships', and who delegated responsibility to lord lieutenants in each county. All cargo was divided into four categories: flotsam – cargo that floated; jetsam – jettisoned cargo abandoned by the crew in their attempts to save the ship; ligan – cargo that sank and was marked with a buoy for later retrieval, and wreck – the cargo that was washed ashore. For many years, all four types became the subject of an undignified wrestle between the finder, the landowner, and the Crown. If the rightful owner did not claim his cargo within a year and a day, it was forfeit to the Crown, although the finder was entitled to a reward proportionate to the value of the goods.

Landowners could claim the 'privilege of right' to anything washed up on their foreshores. Their tenants then adapted that privilege to suit themselves. Wreckers took advantage of the silences in legislation to justify their lootings either under civil law, or under a shrewd interpretation of divine justice. Eventually, the impasse developed into a very British truce, part opportunism, part Queensberry Rules, and part amateur criminality. With the increase in customs and smuggling patrols during the early nineteenth century, wreckers realised that their safest chance lay in staying within the law; if they came across a stricken ship, they must rescue the crew first, but, having done so, the ship became theirs to plunder or sell as they pleased. The practice still exists to this day; any ship (other than those of the coastguard or RNLI) that assists another ship in distress is entitled to claim a portion of the value of that ship in return for saving the lives of the crew. Given that a captain therefore risks forfeiting his ship, this also gives rise to the reluctance of many crews to issue Mayday calls, even in extremis.

The early wreckers also brought a certain grim ingenuity to their tasks. Many locals in areas in which ships were regularly wrecked did not just wait for disaster; they created it. Luring ships onto the rocks was a particular favourite. The Scilly Isles, the West Country and the Hebrides were all rumoured to have wreckers who put up false lights to guide the mariner onto the rocks. It was easy enough to light a bonfire on a dangerous coastline, or tie a lantern to a horse's tail so it imitated the swinging of a ship's light. For a while, the first lighthouses only made the situation worse. The local wreckers, aware that ships relied on the towers to know their position near land, set up rival lights nearby in order to beguile the pilots away from their true course and onto the nearby coast. There were other methods as well. The Wolf Rock, eight miles south-west of Land's End, was a notorious hazard for shipping, and was regarded by the local Cornish wreckers as an excellent source of plunder. Within the rock, however, there was a cavern hollowed out over centuries by the movement of the tides. When the waves crashed through it, trapping and then releasing the air within, the cavern made a sound eerily similar to a wolf's howl. The wreckers, worried that the lonely baying of the rock would alert ships to the Wolf's existence, stopped up the cavern with stones to silence it.

Unfortunately, the Scots were no kinder. Compton Mackenzie's amiable fable of the *SS Politician* in *Whisky Galore* was based on a less amiable truth; the Highlanders and Islanders of Scotland were enthusiastic wreckers. Legends and rumours seeded themselves with suspicious frequency; the local minister on the Isle of Sanday was reputed to pray devotedly every Sunday for those in peril on the sea, to ask God politely if he intended to sink any ships soon and, if so, whether He couldn't organise it so they were wrecked on Sanday. When Robert Stevenson started work on the island in 1806, he noted that wrecks were so frequent in the area that the islanders fenced their fields with ship-timbers instead of stone. Wrecking also produced another curious inequality; rents on the sides of the island that produced most wrecks were higher than on the more hospitable side. Living in a wreck zone had kept the northerners rich, and the southerners poor. Robert was also astonished to discover 'a park paled round, chiefly with cedar wood and mahogany from the wreck of a Honduras built ship; and in one island, after the wreck of a ship laden with wine, the inhabitants have been known to take claret to their barley meal porridge, instead of their usual beverage.' Thomas – and Robert in his turn – had a hard task in selling their lights to the islanders before they had even begun to build them.

But for all the predictable and unpredictable human difficulties, Smith's early efforts with the Scottish lighthouses provided a useful guide for all his professional successors. He was, after all, not a trained engineer in the modern sense, but an imaginative man who did his best with the materials available. The Commissioners had only a vague idea of what the work would entail, and expected Smith to complete most of the supervision on his own and unpaid. For almost ten years, Thomas took no salary at all from the NLB (who were, in any case, broke) and relied entirely on his income from the Edinburgh work. There was some method in his madness.

Thomas worked for the Commissioners because he believed implicitly in the need for guidance at sea, not because he thought it might profit him. He had been reared with a strong notion of public duty, and was quite prepared, despite the lack of money and the spartan conditions, to live up to his promises. Despite the improvised nature of the work, his reports show a good-natured stoicism for the endless hardships he put up with. He noted everything, from the supply of window putty to the problems the keepers had with grazing for their cows. Where routine could be imposed, Thomas tried; he wrote reports, revised instructions, built relationships and imposed discipline. Once it became evident that lighthouse work would demand an ever-increasing amount of time and attention, Thomas resigned himself to regular annual voyages around the coast inspecting existing lights and assessing the necessity for new ones. The voyages were usually hard and often frustrating; Thomas settled into a familiar pattern of remaining storm-stayed in port or being delayed by the unwelcome attention of press gangs.

When back in Edinburgh, Thomas spent much of his time planning improvements to the lights. There were also the demands of Edinburgh society; Thomas, as entrepreneur and public servant, slid happily into the comforts of the New Town bourgeoisie. He trusted implicitly in the Edinburgh virtues of thrift, hard work, humanity and humbug. In middle age, he grew a little stout, but never idle. He worked hard for his business, looked after his family, and took to holding dinner parties. His make-do background had some influence on his later character; once the business was healthy enough, he became the most conservative of men, joined the Edinburgh Spearmen (a volunteer regiment ostensibly called up to fight the revolutionary French but actually dedicated to suppressing domestic riots) and became a captain. The discipline of his public life coincided nicely with his professional existence. He did well from the New Town, which provided an almost inexhaustible demand for brassware, grates and fittings of all kinds, and fitted into the new middle-class world of salons and afternoon teas with ease.

Thomas had been able both to exploit the new, hubristic mood of the city, and to appropriate many of its values. And, having earned his place in society, he was a contented man. He had overcome great insecurity to establish himself in a role which demanded exceptional effort, but rewarded him with both position and respect. His marriage to Jean Lillie had given him a warm and stable family life, and the lighthouses provided the means to keep it. By 1803, he had been confident enough to buy himself a patch of land in Baxter's Place in the lee of Calton Hill, and to build on it a grand new family house in delightfully fashionable style. It was large enough, indeed, to allow both for a warehouse in which he could experiment with designs, and for a separate flat in which the older children would later be installed. Inside its newfangled elegances, the Smith and Stevenson children lived in disciplined harmony, apparently quite content with the splicing together of the two families. And, it was rapidly becoming evident, his marriage had also gained him an apprentice who seemed to have every intention of continuing his connection with the Northern Lights.

By the age of sixteen, Robert Stevenson had already become an adult. In youth, he appeared a sturdy, rounded young man, with a complexion ruddied by outdoor work and with a deceptive spark of humour in his eyes. He remained uneasy with books and culture, but was completely at home with the practicalities of stone, iron, brass and wood. While at home, he became the model of a conscientious gentleman, attentive to his mother and devoted to his stepbrothers and sisters. He was also becoming a plausible successor to Thomas as head of the family. Even then, he had already shouldered all the adult responsibilities of his future life and was busily developing an ambition to move on in the world. He, like the rest of the Smith-Stevenson brood, felt the need 'to gather wealth, to rise in society, to leave their descendants higher themselves, to be (in some sense) among the founders of families', as his grandson Louis later put it. Above all, Robert wanted to be useful.

Much of Robert's later attitude to life was marked by the experience of his childhood. His early years had shown him first the impoverishment caused by his father's early death, and then, through the move to Edinburgh and his mother's marriage to Thomas, the evidence that merit and enterprise earned their rewards. Above all, they had taught him to trust in himself. He also remained mindful of the sacrifices Jean Lillie had made for him, acknowledging many years later that 'My mother's ingenuous and gentle spirit amidst all her difficulties never failed her. She still relied on the providence of God, though sometimes, in the recollection of her father's house and her younger days, she remarked that the ways of Providence were often dark to us.' The move to Edinburgh and the uniting of the two households had also proved helpful. Thomas's example in iron-mongery and lighthouses had not only settled Robert in his chosen vocation but allowed him to repay what he felt were some of his early debts in life. He was also lucky in his choice. Engineering suited him, drawing out both his fondness for adventure and his talent for mathematical abstractions. It allowed him to be creative, and to contribute something of worth to posterity. Above all, it was a useful, manly sort of trade, requiring both solidity and self-confidence.

For the moment, however, Robert was still preoccupied with the slow climb up the foothills of his profession. During the 1790s, he was despatched to Glasgow University to learn civil engineering under the supervision of Professor John Anderson. 'Jolly Jack Phosphorous', as Anderson was known, was rare among eighteenth-century tutors for being as enthusiastic about the practical applications of engineering as he was about its theory. It was said that Anderson had first interested James Watt in steam power, and, scandalously, that his university classes were based as much on fieldwork as they were on black-board studies. He later bequeathed money to a separate tech-nical college in Glasgow staffed with tutors who would not 'be permitted, as in some other Colleges, to be Drones or Triflers, Drunkards or negligent in any manner of way'. The college flourished, and was eventually to become Strathclyde Uni-versity.

In addition to his classes in mathematics, natural philosophy (physics), drawing, and mechanics, Robert learned much of direct value to Thomas's business, and in later years became an ardent supporter of Anderson's methods. 'It was the practice of Professor Anderson kindly to befriend and forward the views of his pupils,' he wrote later, 'and his attention to me during the few years I had the pleasure of being known to him was of a very marked kind, for he directed my attention to various pursuits with the view to my coming forward as an engineer.'

Having discovered the attractions of a subject he wanted to learn, Robert had also become a keen preacher for the benefits of a sound education. The first fees he earned for his engineering work were passed on almost instantly to his old school, and his letters home are peppered with references to the usefulness of his university classes. Once converted to anything, Robert was always the most fanatic of believers.

Robert also showed an enthusiastic interest in the lighthouses. The mutable quality of the work suited him and after accompanying Thomas on a couple of his regular inspection tours, Robert began to appropriate small patches of lighthouse territory for himself. Thomas introduced him to the Commissioners, allowed him to fit lenses or supervise building work and encouraged him to develop his interest as warmly as possible. By the mid 1790s, Robert appears often in the NLB's Minute books, first as understudy, and then in more significant roles. He already had a sound grasp of all aspects of the business from the sizing of lamps to the sculpting of reflectors. His chief fault, if any, was a forcefulness in his dealings that did not always endear him to potential customers. Within six years of joining Thomas's workshop he was regarded as an equal in almost all aspects of the work, and by 1800 had been made a full partner in the firm.

And so, in the pattern that was to become settled for the next three Stevenson generations, Robert spent his winters at home in the south studying and his summers in the north supervising the details of work on the lights. Much of his education was also completed in Thomas's workshops first at Bristo Street and then at Baxter's Place, making grates for the gentry and lamplights for the Corporation. As master and pupil, Thomas and he were well suited to each other. It was in some ways an odd partnership; Thomas was, after all, not only Robert's employer, but also his stepfather. Stretched too far, the relationship could have become awkward or imbalanced, but as it was, the two made ideal accomplices. Thomas, though a milder character, was a generous teacher. The two men were alike in many respects. Both had been reared the hard way; both believed in the benefits of a stern apprenticeship, and neither took anything for granted. Before he died, Thomas was to realise that Robert's talents would one day far eclipse his own. It is a measure of Thomas's generosity that, far from resenting his stepson's advancement, he was delighted.

THREE

Eddystone

Edinburgh in the early 1800s was an enticing place for a middle-class merchant with an enthusiasm for self-advancement. After the blow administered to its pride by the Act of Union in 1707, the city had descended into a long sulk. Union had allowed Scotland to keep a few trophies of her independence – her own law, a separate educational system, a bishop-stripped church – but the old, bitter quarrel with London did not cease overnight. Even Defoe, whose role as an English agent was to sell Union to the Scots, had found relations difficult. 'Never two nations that had so much affinity in circumstances, have had such inveteracy and aversion to one another in their blood,' he wrote disconsolately. By the 1750s, as trade improved and the benefits of Union began to be felt, the city emerged from its self-absorption. Gradually, it began to look to England and London as its example; there was much talk of Britishness and the first furtive attempts at English speech, English habits and English thinking. Soirees (pronounced 'sorries' in Edinburgh and 'swurries' in Glasgow) were held more frequently, tea was drunk and Scots began, as the philosopher David Hume put it, to be considered, 'a very corrupt dialect'. Fifty years later, Lord Cockburn (rivalled only by Sir Walter Scott for his domination of the Edinburgh scene) wrote gloomily that 'When I was a boy, no Englishman could have addressed the Edinburgh populace without making them stare, and probably laugh. We looked at an English boy at the High School as a ludicrous and incomprehensible monster. Now these monsters are so common that they are no monsters at all.'

The Scottish Enlightenment emerged slowly from this half-derelict background. The great upswell of enterprise and industrialisation produced an extraordinary group of men who came from lowly backgrounds to fill the sudden need for innovation. In the century after Union, Scotland produced an exceptional group of artists, philosophers and scientists, including Burns, Smollett, Adam Smith, Alan Ramsay, Robert, William and John Adam, Walter Scott, James Hogg, Henry Raeburn, James Watt, Thomas Telford and David Hume. The men who guided the Enlightenment were united by a growing belief in the force of reason. Man, they argued, was no longer at the will of his environment; he could explain it, control it and shape it where necessary. Life in all its aspects could be improved upon; there was to be no such thing as an old truth. Faith could be questioned, landscape could be shaped, economies could be transformed. In particular, they came to regard the study of mankind and the improvement of human nature itself as an essential part of enlightened life. They put aside the unreasoning faiths of the pre-Union years and replaced them with a new philosophy, brisk and scientific. Scotland was no longer forcibly strapped to her past; it was possible to improve on history.

The intellectual adventurousness of the age was matched by a rush of enterprise. Agriculture and business flourished, new industries boomed and old practices vanished. The expansive mood was a blessing for entrepreneurs. Many of the richer merchants took to the ultra-fashionable new hobby of agricultural improvement, turning their acreages into models of economic discipline and landscaped tameness. Some went north to enlighten the misguided Highlanders. They planted trees, started enterprises and encouraged the use of new machinery. Alongside the new flocks of scientists and social engineers, the inventors throve. Engineering, previously considered a profession for tradesmen and foreigners, began to develop a status and confidence of its own. Some of the mill-wrights, masons and clerks who moved into the profession were shrewd enough to see the urgent need for new design, and rose to meet the challenge. Down in England, Smeaton was building lighthouses, Arkwright was showing off the benefits of water frames and Trevithick was designing prototype locomotives. In Glasgow, James Watt moved on from making musical instruments to experimenting with steam engines. In the north, Thomas Telford had begun threading roads around the coasts, while John Rennie was building bridges. By the mid-nineteenth century, Prince Albert was heard to note approvingly that 'If we want any work done of an unusual character, and send for an architect, he hesitates, debates, trifles: we send for an engineer, and he *does it*.' The heroes of post-Enlightenment Britain also encouraged the view that hard work, imagination and enterprise were all that a man needed to rise from the lowest level of society to the highest. Titles did not matter quite as much as ability. It was small wonder that Thomas Smith and, in his turn, Robert Stevenson felt themselves in familiar company.

Edinburgh's architects, meanwhile, were building the New Town. The city that Defoe had visited in the 1720s was, by even the most optimistic standards, disgusting. Overcrowding, disease and squalor had given Auld Reekie its name and reputation; Glasgow, by comparison, was the finest, sweetest city in all the Empire. The lack of housing and the density of people meant that people took shelter where they could find it. Each of the tottering Old Town tenements housed a cross-section of every class and occupation from Lords to barrow boys. The Proposals for Carrying on Certain Public Works in the City of Edinburgh were drawn up by the Convention of Royal Burghs in 1752 and construction work began soon after. The Nor' Loch below the Mound was drained and landscaped, the Lang Dykes became Princes Street and the slow geometry of the Georgian New Town began to unfold. Work had not been finished before the middle classes bolted from their squalid quarterings near the Castle to the new city.

The division between the old town and the new was the most eloquent illustration of the divisions in Edinburgh's character. It had always seemed the most strong-minded of all Scots cities, but under the surface the contradictions became more obvious. During the eighteenth and early nineteenth centuries, it managed to sustain several wildly contradictory faiths: anti-Englishness and fervent Britishness; improvement and nostalgia; depression and vivacity. It never did, as it sometimes liked to believe, exist in cosmopolitan isolation. During this period it feared the invasion of the French, the Papists or the Wild Highlanders even more than it feared the loss of its identity to England. The terror of anarchy produced a contrariness in the city's character, at once devout and cruel, reasoned and unreasonable.

Both the Smiths and the Stevensons had come to Edinburgh from elsewhere, but ended up adopting the habits and thoughts of the city as their own. By the turn of the century, they had become perfect exemplars of New Town life. Baxter's Place had become a useful port of call for clients who had heard of Thomas and Robert's growing reputation for ironmongery, lamp-making and general engineering projects. Within the house itself, life became segregated; while the men discussed work, the women gossiped over acquaintances and fussed over the children. The link between the two families had been reinforced by Robert's wedding in 1797 to Thomas's eldest daughter, Jean (one of the two surviving children born to his first wife). 'The marriage of a man of 27 and a girl of 20 who have lived for twelve years as brother and sister, is difficult to conceive,' Louis later commented drily. But Thomas much approved of the union, regarding it as another healthy symptom of the closeness of both families. Outsiders, however, regarded Thomas's status as Robert's stepfather, boss, father-in-law and mentor as dangerously intimate.

Something of the strangeness of this arrangement comes over in Robert's letters, which mention in the same breath his half sister Betsey, aged six (Thomas's daughter by his marriage to

Robert's mother) and his wife, Jean, Thomas's daughter from his previous marriage. In a letter of 1807 sent to Robert while he was aboard the lighthouse yacht, Thomas signs himself 'your ever affectionate father', and mentions that 'we are all well: Your wife, your mother and myself dined and drunk tea with Mr Gray yesterday.' In letters and diaries, Robert would refer to Mary Anne, Thomas's daughter from his second marriage, as 'my dear sister', and later regarded the offspring of the Smith children not as unrelated acquaintances, but as cousins. As Thomas grew older, Robert slipped naturally into position as head of the family, presiding over both Thomas's children and, increasingly, his own. Robert's marriage was certainly a curious match, even when judged by the habits of the day. But the two families had been living together for over a decade, Robert had learned his trade, his morality and his habits from the Smiths, and, in the enclosed New Town world of the 1790s, such unions were not unusual. Besides, Robert's mother seems to have done a little engineering herself, and was no doubt delighted at the wedding.

In common with the rest of her family, Jean Smith possessed a forceful character. As befitting her respectable position, she was a refined and rather feminine creature, who had for the most part led a life of bourgeois niceties. In her way, however, she was as single-minded as her new husband. She ran a well-disciplined household, founded on firm government and Christian virtues, but at the same time remained interested in high society. Louis, it seems, found Jean's charms difficult to understand. 'My grandmother,' he wrote, 'remained to the end devout and unambitious, occupied with her Bible, her children and her house; easily shocked and associating largely with a clique of godly parasites'. As someone who had struggled with religion for much of his life, he found Jean Stevenson's sheeplike devotion exasperating. If she was asked to employ a cook, the cook would be taken because she was pious, not skilful. The midwife would be hired on her knowledge of the Catechism rather than her gift for obstetrics. Louis was hard pressed to understand what it was about her that Robert had found so compelling and wondered slyly about 'the sense of disproportion between the warmth of the adoration felt and the nature of the woman'. He searched a little further, and came up with only the faintest of praise. 'She diligently read and marked her Bible; she was a tender nurse; she had a sense of humour under strong control.' Robert might have been allowed to play the adventurer away from home but once returned to Baxter's Place he deferred to his wife. He did not apply the same standards to marriage as he did to work; women, he believed, were a softer, more obscure species than men, and could not be dealt with in the same blunt manner as men. In belief and habit Robert was an old-fashioned man and Jean an old-fashioned wife. Despite this, or perhaps because of it, they had an equitable marriage. Robert remained to the end devotedly loyal and prepared to respect Jean's judgement in almost all family matters. The only household business which he took an interest in was the education of his children; the rest was her responsibility.

Jean was also marked by the deaths of her children. Between 1801 and 1818, nine children were born, of whom five survived. 'Never,' wrote Louis with the blithe callousness of a born survivor, 'was there such a massacre of the innocents; teething and chin-cough and scarlet fever and small-pox ran the round; and little Lillies, and Smiths, and Stevensons fell like moths about a candle.' By 1818, the carnage had ended and the Stevensons were left with Jane, Alan, Bob, David and Thomas. But Jean reacted to the loss of the four souls by lavishing a morbid attention on the remaining children. Alan, in particular, worried her; he was a pale, frail child whose childhood was often interrupted by illness. Jean pestered him endlessly to wrap up and to keep taking the poultices, and Alan treated her fussings with a mixture of embarrassment and dismissive gratitude. Jean's melancholic habits may be one reason why the five surviving Stevenson children became so superstitious about illness. All of them grew up obsessively attentive to their own wellbeing; Thomas, in particular, became a full-time hypochondriac. The Stevenson domestic life, and the ritual outbreaks of Edinburgh epidemics – cholera, tuberculosis and smallpox were all, at various times, rife – made a potent mixture. The contrast between the hardiness of their lighthouse work and the dainty medical paranoia at Baxter's Place was to become even more emphatic in later years.

Several consolatory letters from his staff reveal the extent that the deaths of his children preyed on Robert's mind at the time. His letters home became more and more solicitous, enquiring often about Jean's health and the state of the remaining children. He hoped Jean was getting out enough, that she was taking plenty of exercise, that she saw her friends often, that she went regularly to market. In 1816, after yet another death, he advised her not to become too depressed. 'If that kind of sympathy and pleasing melancholy, which is familiar to us under distress, be much indulged, it becomes habitual, and takes such a hold of the mind as to absorb all the other affections, and unfit us for the duties and proper enjoyments of life. Resignation sinks into a kind of peevish discontent.' His concern for her was also transmitted to the children. They were to attend to their studies, get plenty of fresh air and not to dwell on morbid thoughts. 'Let them,' he wrote to Jean, 'have strawberries on Saturdays.' He also chivvied the children about their education. Jane, his eldest child, was instructed to 'read Wotherspoon, or some other suitable and instructive book', Bob to 'learn his Latin lessons daily; he may however, read English in company'. While in Fraserburgh, he suggested that 'it will be a good exercise in geography for the young folks to trace my course'. Above all, as he wrote in one communal letter from London, 'the way to get money is, become clever men and men of education by being good scholars.'

The Smiths also remained close. Jean's sister Mary Anne (always known as Mary) remained at Baxter's Place long after

Robert had taken over from Thomas as patriarch. She played an active part in bringing up Robert's children and, since she remained unmarried, became almost a second mother to them. James, meanwhile, moved away from Baxter's Place to found his own ironmongery business and establish a family. Mary and Robert got on well, though in later years Robert tended to take the tone of overbearing elder brother with her, and became peevish about the burden she placed on the household expenses.

Poor Mary, amiable and sheltered, spent much of her time shuttling between Robert and Jean, caught in the unenviable trap of Victorian spinsterhood. Later, she was to try taking a position as a governess in London; her letters from Islington show a bewildered innocence about life beyond Edinburgh and a terror of displeasing her stepbrother. She wrote to Robert in 1818, informing him with wary cheerfulness that she had been sightseeing at the Missionary Museum for Heathen Idols and the Houses of Parliament where she 'had the misfortune to trip over the Wool Sack'. She had also been on a trip to Hampstead, which had left her most disappointed at the dullness of English scenery. English prices were terrible; the boarding house she was staying in had already cost fifty guineas. 'Should it be found necessary that I should do something for myself, I will prefer a situation in England for some time,' she added disconsolately, evidently aware of Robert's ferocious attitude to laggards. 'Your letters contain both instruction and amusement . . . I shall also obey your injunctions to attend the Established Church.'

Thomas, meanwhile, was becoming exhausted by the annual circuits of Scotland. He spent more time concentrating on his lamp-making business, finding it less demanding than the lighthouse work. By 1800, having reached something of the status of a Grand Old Man, Thomas handed all the lighthouse business on to Robert and retired to a respectable dotage. Robert still referred to him on details, and kept up regular reports of his progress while away, but by 1808 had separated both his lighthouse and private business from Thomas's old company. The two remained close in business and private, and Robert found it reassuring to have his stepfather's guidance for his schemes. But by 1800, Robert had already outstripped Thomas's experience, and was becoming a recognised expert in his own right. By 1789, the NLB had been given a free hand to build lights when and where they felt they were necessary, and between 1793 and 1806, five more were built at points around the Scottish coast. The NLB scarcely had to bother looking for a successor to Thomas; Robert was appointed as engineer to the Board as naturally as if he had been born to the role.

The outdoor life of a lighthouse pioneer suited Robert wonderfully. Thomas had put up with the voyages, obstacles and disputes; Robert actually enjoyed them. He, like Thomas, lived a sober life in Edinburgh, but once released from the city, his character expanded. His job, after all, was an unusual one. As his grandson Louis later noted,

> The seas into which his labours carried the new engineer were still scarce charted, the coasts still dark; his way on shore was often far beyond the convenience of any road; the isles in which he must sojourn were still partly savage. He must toss much in boats; he must often adventure on horseback by the dubious bridle-track through unfrequented wildernesses; he must sometimes plant his lighthouse in the very camp of wreckers; and he was continually enforced to the vicissitudes of out-door life. The joy of my grandfather in this career was strong as the love of woman. It lasted him through youth and manhood, it burned strong in age, and at the approach of death his last yearning was to renew these loved experiences.

Robert spent the next few years supervising the lights and correcting any existing problems. Like Thomas, he was meticulous, but he brought a blunter approach to the job than his predecessor. Perhaps his greatest asset was the force of personality necessary to carry a reluctant staff along with him. As it became evident that the lighthouse work was swallowing more and more of his time, he began training assistants to help. In the early days he still did the bulk of the work entirely alone, but as time went on, he began to build relationships with a select group of workmen whom he felt he could trust with many of the mundane details. The assistants, in turn, understood Robert to be an employer who expected backbreaking work but offered in return loyalty and enlightened working practices. But it was not merely a decent wage that kept many of his assistants with him for the whole of their working lives. Robert had charisma; the charisma of great purpose. He was not charming, and would have considered charm to be an insult, but he did have the ability to lead others through sheer force of personality. He persuaded not through words but through deeds and by his own example.

At times, Robert's intensity could be daunting. In many cases, he silenced suspicion or opposition by simply ignoring all argument on the subject. Anyone who complained that lighthouse work was too tough or too dangerous had only to watch Robert sailing through gales or striding across wave-swept reefs to feel themselves fainthearts in comparison. Anything his workmen could do, Robert himself would prove he could do better. For all his high-handed habits, Robert did not regard himself as a grandee but as an equal partner to his men. His flaw was to expect exactly the same of others as he expected of himself. As Louis later put it, 'what he felt himself, he continued to attribute to all around him.' Robert was a worker and an impassioned preacher for the benefits of self-discipline, and those who fell below his particular standards astonished him. When others let him down, as they often did, Robert reacted with bafflement; if he was fair with them, he reasoned, why were they not equally straight with him? Though a great leader, Robert was too often a hopeless judge of human nature. He knew himself, and considered that knowledge sufficient.

For much of the time, Robert was worrying about theory as well as practice: would a revolving light suit one place better than another; were floating lights a feasible option; could the cost and difficulty of transporting oil be offset with other savings? Robert set about refining Thomas's original designs for the reflectors, applying a silver coating to the old copper circles, and dispensing with the beautiful but ineffective silvered-mirror designs. His refinements helped to increase the strength of the light, and, when several reflectors were placed one above another in the light room, gave out a steadier, better light than Thomas's weak beams. Some time later Robert was also flattered to discover that the reputation of his new silvered reflectors had drifted far beyond Scotland. The Covent Garden Theatre in London ordered one and spent some months experimenting with it as a potential spotlight. John Sam, the theatre carpenter, eventually returned it to Robert, mentioning regretfully that 'It is an excellent reflector but it collects the light too much in one spot for our use.' Theatre's loss was the lighthouse's gain, and Robert was delighted with this unintended endorsement of his designs.

He then began concentrating on the oil lamps themselves. Twenty years previously, a Swiss scientist named Ami Argand had begun experimenting with the idea of an oil lamp which could give a purer, brighter light than the smoggy lamps then in use. As Argand's brother later told the tale, he and Argand were eating supper one night when he broke the neck of a glass flask. Reaching over to pick it up, he accidentally moved it over the flame of the oil lamp. 'Immediately it rose with brilliancy,' he wrote. 'My brother started from his seat in ecstasy, rushed upon me with a transport of joy and embraced me with rapture.' With that broken flask – later developed into an elegant glass chimney – Argand had discovered a method of protecting and clarifying the flame from an oil lamp. Once he had added a circular wick, and a lever to raise or lower it as necessary, his invention became popular throughout Europe and beyond.

Robert, realising how useful the invention was for his purposes, tested a prototype at Inchkeith and then brought in Argand lamps for all the Scottish lights. It was, he discovered, a sharper and more reliable light than Thomas's original lamps. Depending on the oil used, the strength of the beam could also be increased or decreased. At various stages, Robert experimented with olive and rapeseed (or colza) oil, and tried to find alternatives to the smoggy household oil then in use. He also read of a South American discovery that sheep's tails produced an unusually brilliant light. Unfortunately, the difficulties involved in clipping the tails off thousands of reluctant sheep were found to be insuperable. Whale oil seemed the best option for Robert's purposes; it was expensive but effective, burned cleanly and gave out the brightest light. Sperm whales, already heavily hunted in both northern and American waters, therefore provided the Scottish lighthouses with their fire for another fifty years.

With the increase in lighthouses came the need to differentiate between lights. In the Pentland Firth, for instance, where several lights would eventually be visible at one time, it was necessary to distinguish each with an individual pattern of flashes. At the time, all lights were fixed, sending out a steady beam in one direction throughout the night. Robert, pondering how best to vary the lights, began slowly to devise methods of rotating the reflectors on a central axis so they appeared instead to flash. His first few experiments were troublesome, since it seemed almost impossible to devise a smoothly balanced mechanism for the cumbersome ranks of lamps. In practice, it often required one of the keepers to spend the night pushing the light round in circles. Nevertheless, by 1806, the new lighthouse at Start Point included a clockwork mechanism 'exhibiting a brilliant light once in every minute, and becoming gradually less luminous'.

Lenses, of course, still did not exist for lighthouses, though some experiments had been undertaken in England and France with pieces of glass placed behind the light. All lighthouse technology developed in painstaking isolation. Those researchers who were experimenting with new engineering or optical techniques were spread far away around the laboratories of Europe, and there was as yet no way of communicating their findings efficiently. Even by 1819, it was calculated that, out of 254 lighthouses then in use throughout Europe, 5 were still lit by candles, 30 by wood or coal fires, 2 with coal gas, 157 with common oil lamps and 60 with Argand lamps. Many countries were not aware of developments or inventions elsewhere, and, though the different national lighthouse services often worked well together, most stuck resolutely to their own practices. Indeed, for a while the NLB had rather better relationships with the French lighthouse commissioners than it did with the English.

In practice this meant that Robert often worked blind, unaware of the efforts of his fellow engineers and unable to do more than feel his way through the darkness. In time, Robert established links with many of the most important British and European inventors and an atmosphere of mutual helpfulness, jolted by the occasional hiccup of professional jealousy, gradually developed. For the moment though, Robert was more preoccupied with refining his existing designs. He spent his winter nights making elegant architectural drawings, detailing the dimensions of each lantern, and polishing his reflectors to perfection.

Robert relished the journeys and fretted at the endless delays spent in harbours or waiting for supplies. Behind the occasional non-committal statement in his diary or the NLB Minutes that he 'found the light in good order', lay hours of patient pedantry. His written evidence reveals as much by its omissions as by the endless records of missions accomplished. On paper, Robert was not an eloquent man. His diaries and minutes tell a gruff history only occasionally illuminated by flashes of insight. His grammar, as Louis later noted with some exasperation, was often hopelessly tangled and his occasional attempts at flattery ill-judged. As he wrote offhandedly to a colleague in 1802, 'In submitting this address to you, I was otherwise impressed than with the view of laying something before you that might afford pleasure from style and composition. Those are no happy talents of mine, for my avocations in life intirely preclude me from such advantages.' His son Alan was later to find Robert's habit of referring to himself in official correspondence as 'the Reporter' or 'the Writer' mystifying. But Robert, unlike his son, remained conscious throughout his life of the gaps in his education. He spent twelve years chasing his degree, and was finally prevented by his lack of Greek or Latin. His solution was to turn their absence into an advantage, and to be as plain-spoken as possible in all his dealings. Sometimes, indeed, his brusqueness verged on bullying. In all his work there was a sense of barely stifled urgency; he wanted life to be as fast and as efficient as he was, and he grew almost frantic when transport, communication or human fallibility disappointed him.

But along with his desperation for haste Robert had extraordinary patience for detail. Most engineers, planning out a bridge or a road or a light, would have given only the most necessary sketchings. Robert gave ornaments or cornicings the same attention as he did load-bearing walls. He didn't need to bother; he just wanted to. Likewise, he remained meticulous about the regular blizzards of paperwork that the lighthouse service produced. He would often return late at night from a day's hard journeying and start on a further mileage of instructions. At one stage, he calculated, he had written and received over 3,000 letters in a single year, in addition to preparing the rough and fair copies of reports, estimates, invoices and private correspondence. For a while, he also kept up a separate journal, a memorandum book and a diary. When his youngest son, Thomas, visited England in 1844, Robert handed him a crowded little notebook containing jottings on almost every harbour in Britain. Again, Robert had not written it from necessity, but from an instinctive thoroughness in all his dealings, however apparently irrelevant.

As chief engineer, Robert was also expected to contend with less predictable difficulties. One of the major hazards of any journey around the Scottish coast at the time were the press gangs. The boom in trade and war and the premium it put on able seamen meant that shipowners often had to find more inspired methods of attracting a crew. At the end of long voyages, ships would frequently find themselves surrounded and boarded by gangs intent on kidnapping or coercing the sailors into work on other boats. Sometimes they acted for themselves; often they acted for the State. The navy, with its voracious hunger for manpower, had become the press gangs' best customers by the time of the outbreak of the Napoleonic Wars. The pressmen used every method, legal and illegal, at their disposal to find sailors, from searching the workhouses to plundering the country's prisons. Most gangs relied on a steady supply of informers, and were prepared to consider almost anyone, including deserters, criminals, children and the disabled. An Act of 1704 declared that 'Idle Persons, Rogues, Vagabonds, and Sturdy Beggars . . . are hereby directed to be taken up, sent, conducted, and conveyed into Her Majesty's service at sea.' Later amendments to the Act provided for a suit of clothing to be given as a bribe to volunteers, though it made little difference, since the victims of impressment rarely went quietly.

The sailors, aware of the charmless fate that awaited them, took as much evasive action as possible. Some rioted, some capsized the press gang boats, and some feigned idiocy or injury (burning a fresh wound with vitriol to make it look like scurvy was, for a while, a favoured practice). Even when ashore, competent mariners were unsafe; press gangs patrolled the major maritime cities in search of recruits, while epidemics swept the docks. Fit and healthy men fled inland, or deserted as soon as they came on shore. By 1810, the Royal Navy employed around 150,000 men, many of whom, as one observer noted, were 'so much disabled by sickness, death and desertion' it was a miracle they sailed at all.

From Robert's viewpoint, the press gangs were at best an inconvenience and at worst an active hazard. Most troublesome of all was the gangs' habit of taking those who were, in theory at least, exempt from impressment. For a while, Orcadians were excluded from the press gangs' depredations since they argued that their work on the sea was essential for the survival of the islands. But as the hunger for men increased, they too often found themselves duped into service on the boats. Robert, who often hired builders or sailors from Orkney, swiftly discovered that special pleading made not the smallest difference. On a number of occasions, he was forced to lurk several miles out from harbour to avoid the acquisitive habits of the press gangs. On another occasion, his sailors were only saved by the thoughtfulness of a female passenger on the lighthouse yacht. 'With much fortitude and presence of mind,' he wrote later in his diary, 'she offered to conceal them under the state room bed, she lying on top and feigning illness. This plan succeeded so well that the affair was never suspected and the men got clear.' The Stevensons found the practice of impressment exasperating, but no matter how much they pleaded that their men were needed for work on the lights, the gangs went on taking their quota time after time.

Nor was Robert exempt from the attention of the wreckers. When he complained to one of the local fishermen on Sanday about the conditions of his boat's sails, the man replied slyly that if Robert hadn't brought his lights, then all the islanders might have had better sails, better boats and a better life. Visiting mariners might have welcomed their efforts; the locals usually hated them. As Robert reported to John Gray, the Clerk of the NLB, in 1802, 'You would hardly believe with what an evil eye the Wreck Brokers of Sanday view any improvement upon this coast, and how openly they regret it. . . . You will readily see . . . how deeply their interest is concerned when

I can assure you that since the erection of North Ronaldsay lighthouse, a period of about 12 years, upwards of Twenty Vessels have been wrecked upon the island of Sanday.' While on one of the lighthouse inspection voyages many years later, Robert and his young son Thomas (father of Robert Louis Stevenson) were making their way through the Pentland Firth when fog came down and the crew dropped anchor for the night. When they woke at dawn the next morning, they discovered that the ship had drifted close to the isle of Swona. As the mist rose, they found themselves staring at a broad sandy bay, and above it, a small hamlet of fishermen's huts with all the occupants apparently still asleep. If the current pulled them any closer to the island, the ship would have run aground, so the captain fired a gun as a distress signal. One by one, the villagers emerged from their huts, and stared across the beach at the drifting ship. There was a long, slow silence. Louis took up his grandfather's account. 'There was no emotion, no animation, it scarce seemed any interest; not a hand was raised; but all callously awaited the harvest of the sea, and their children stood by their side and waited also. To the end of his life, my father remembered that amphitheatre of placid spectators on the beach, and with a special and natural animosity, the boys of his own age.' The ship escaped but Robert's and Thomas's memory of the wreckers remained.

Robert's reaction to the wreckers and pressmen was characteristic. He saw the work he was doing as the errands of public duty. If he ever had a moment's doubt in the need for his work, he never expressed it. Those who opposed him – the wreckers, the locals, the pressmen, and even the sailors themselves – he found inexplicable or downright criminal. He regarded his mission to bring light into darkness as self-evidently justified, and remained bewildered by anyone who saw matters otherwise. 'We have been boarded by the press-gang,' he wrote wearily to John Gray in 1804 while in the lighthouse sloop off Kirkwall, 'we have much of privateers here, but hope should any of them come in our way that they will consider the importance of our mission and let the vessel pass.' It was a forlorn hope. As Robert discovered, not everyone felt as he did, and not everyone could be persuaded by logic, reason or force. Once in a while, he found himself becoming a little cynical. In a letter of 1806 to his newly appointed foreman, Charles Peebles, he gave vent to his frustrations. 'I have the fullest confidence in your candour,' he wrote, 'and that you would use no man ill, but I fear you have been too indulgent on the [men]. I am sorry to add that men do not answer to be too well treated, a circumstance which I have experienced and which you will learn as you go on with business.'

Once the first few lights had been completed around the coast, Robert returned to Edinburgh and the usual winter battles with the Lighthouse Commissioners. They, like Robert, were keen to continue the construction programme, having been petitioned by various town councils for lights in their area. During the long Edinburgh evenings Robert drew up schemes for new lights and kept an eye on Thomas's ironmongery business. His records for the time show a steady flow of reports and estimates for harbours, bridges, piers, canals, drainage schemes, steamboats, roads, memorials, prisons, railways and fog-signals. He spent time drawing up a scheme for heating churches with steam, considered the improvement of Highland roads, and addressed at length the problem of gunpowder storage. He also wrote to local landowners, suggesting the construction of harbours, breakwaters or roads in their areas, and playing heavily on their desire for improvement and prosperity. He might have undertaken the lighthouse work from a strong sense of altruism, but he was also rigorous at maintaining the commercial side of the business.

As an addition to his already burdensome commitments, Robert decided to make several trips to the English lighthouses. They would, he hoped, teach him much about the conditions of the English coast and make an interesting comparison to his work in Scotland. And so, in 1801, he set out for the south. That journey and the two subsequent trips he made in 1813 and 1821, were more revealing than he could have hoped. The lighthouses that he came to inspect had been built on altogether different principles to the Scottish lights. Their development had been a disjointed affair, which provided its own exotic history of heroes and villains. What Robert saw was the culmination of three centuries of work and occasionally misguided effort; and the English lights were to prove useful to the Stevensons not just as templates for their own endeavours but as excellent cautionary tales.

The English lighthouse service started with good intentions but rapidly degenerated into an early example of the perils and benefits of privatisation. While the lighthouses were in the gift of the Crown, their administration was controlled by Trinity House. The governing body, known as the Elder Brethren, had been granted the sole right to build lighthouses in England and Wales since 1514. Unfortunately, its original charter had not given Trinity House any authority to collect dues from shipping, and the Guild was therefore left without funds to build new lights or maintain old ones. It fell back on the obvious solution, and got someone else to do it. For nearly three centuries, all the lights around the English coast were built by individuals who had been granted private patents by Trinity House. The patents, which usually cost only a nominal sum, were repaid through light dues charged to any shipowners using the lights. Lighthouses thus provided a tidy source of funds for many private owners, who had the satisfaction of having achieved something civic-minded at the same time as harvesting a useful income. But apart from granting charters, Trinity House seemed to do everything it possibly could to avoid building lighthouses itself. Between 1600 and 1836, it managed to construct only one light of its own; all the remaining lights around the English coastline were either built privately, or taken over by Trinity House once they had become profitable.

Robert's first English tour was aimed partly as a method of gathering information on the English lights, and partly as a comparison of his own experience with that of the Trinity House men. The Scots service might have been in its infancy, but England's was nearing its dotage, and had much to offer the expert tourist. Robert visited fourteen lighthouses, covered a distance of 2,500 miles and picked up several useful tips for his own work. The differences between the Scottish and English lights, he noted, were often to his, and Scotland's, advantage. The main variations were in materials; the English used timber and metal more than stone and most lights used copper reflectors rather than mirrored ones. From the size of windows and the design of the towers, Robert concluded that the English lights were often less sturdy than the Scots. Costs varied too. The average price of an English reflector and lamp at the time was £1,000. In Scotland it was £600. His account of the trip is characteristic; he showed concern for the standards of English lights alongside a desire to learn and not just to preach. He also disagreed with the practice of allowing individuals to build lighthouses, dismissing the owners as merely 'private adventurers'. In his account, there is also a glow of satisfied competition. In many instances, he found, the lights that he had read so much about did not live up to his expectations. 'I was made fully sensible that the pleasures of anticipation often greatly exceed what is really enjoyed,' he noted.

Not everything on his tour went according to plan. While making enquiries about the Lizard light, 'a young man, accompanied by several idle-looking fellows, came up to me and in a hasty tone said, "Sir, in the King's name, I seize your person and papers."' The authorities, it emerged, believed him to be a French spy.

> The complaint proffered against me was – that I had examined the Longships lighthouse with the most

minute attention, and was no less particular in my
inquiries at the keepers of the lighthouse regarding the
sunk rocks lying off the Land's End and the sets of
the currents and tides along that coast; that I seemed
particularly to regard the situation of the rocks called
the Seven Stones, and regret the loss of a beacon which
the Trinity Board had caused to be fixed upon the Wolf
Rock; that I had taken notes of the bearings of several
sunken rocks ... Further, that I refused the honour of
Lord Edgecombe's invitation to dinner.

Robert produced his letters of introduction and authority and
was told by the local Justice of the Peace that these were 'merely
bits of paper'. For a while, it seemed as if he would be kept in
custody indefinitely until two further Justices cleared him of
any suspicion of espionage and left him to go on his way, 'which
I did with so much eagerness that I gave the two coal lights
upon the Lizard Point only a very transient look'.

The most famous of all English lighthouses, and the one
which most interested Robert, was the Eddystone light. He was
unable to see it in 1801 but returned twelve years later, intent
on examining the light that had provided him with inspiration
for many years. The journey was part research, part pilgrimage;
the Eddystone was built by flawed heroes and, though Robert
did not know it then, was to go through four different designs
before it was finally completed. It was the first light to be built
offshore in Britain and provided each subsequent generation of
engineers with a useful precis of lighthouse design. The history
of its multiple constructions also gave Robert vital guidance for
his future career.

The Eddystone reef lies fourteen miles south-west of Plymouth. Most of it is submerged, with only three feet protruding
from the sea at high tide. The rust-coloured gneiss is as resilient
as diamonds and the currents that surround it send up abrupt
spouts of water on even the calmest days. It is thought of as a

bad-tempered place, full of sulks and strange moods, and by the sixteenth century its reputation for destruction had already spread well beyond Cornwall. Plymouth had every other advantage, including a wide, sheltered bay, a naval dockyard and a vigorous trade with the New World. But merchant captains were so alarmed by the prospect of being wrecked on the Eddystone that they often ran themselves aground on the Channel Islands or the northern French coast trying to avoid it. In 1664, Trinity House was petitioned by two local men for permission to build a lighthouse on the reef. The Elder Brethren rejected the petition, citing the risks involved and complaining, with impressive illogicality, that, since there was no precedent for an offshore light, it must therefore be impossible to build one. Thirty years later, with local pressure mounting, another petition was presented. Trinity House dithered for two further years and then decided that, instead of undertaking the project themselves, they would hand it out to a Plymouth man, 'at his own cost and entire financial risk'. If it worked, they would, after a discreet interval, take the dues; if it didn't, they had lost nothing.

The owner passed the project over to Henry Winstanley, an English eccentric of the finest breed. He was a man of many enthusiasms, an investor, designer, engraver, painter, pamphleteer, illusionist and inspiration for Winstanley's Waterworks in Hyde Park, a show of 'the greatest curiosities in waterworks, the like never performed by any', which ran for over thirty years. Up until 1696, his only connection with engineering or the sea had been as a shipowner. One of his five merchant vessels had already been wrecked on the Eddystone, and a second, the *Constant*, ran aground on the reef before the end of the year. When Winstanley heard of the grounding, he galloped angrily off to Plymouth, demanded an explanation, and was told of the problems in lighting the rock. Winstanley presented himself to the new owner, and told him that he personally would build a lighthouse on the Eddystone. The owner, unconcerned by

Winstanley's lack of experience and wayward reputation, accepted.

Winstanley's design for a light showed all the idiosyncrasy of his previous works. It was built from a combination of iron, wood and brick, stood eighty feet high, and strongly resembled one of W. Heath Robinson's more elaborate contraptions. Above the octagonal brick base, there was an elegant iron balcony, a domed cupola and a glazed lantern, which was to hold 'sixty candles at a time besides a great hanging lamp'. On top of the lantern, Winstanley placed a large and subtly wrought iron weather vane. Even by the standards of the day, it looked extraordinary. On 14 November 1698, the lantern was lit for the first time, Winstanley noting modestly that 'it is finished, and it will stand forever as one of the world's most artistic pieces of work.' Despite this, when Winstanley returned the following spring he discovered that the tower had been damaged by gales. The original mortaring had not set properly, and the resident keepers said they had felt the tower shake beneath them in bad weather. Winstanley decided that the lighthouse needed reinforcement, and began building a new stone casing around the old tower. He also added iron bands around the girth of the base, increased the height to 120 feet, and doubled the ornamentation. The new Eddystone now boasted a new and enlarged gallery, two cranes, several Latin inscriptions, a State Room ('very well carved and painted') and six large candlesticks 'for ornament' on the exterior. Inside there was 'a very fine bedchamber with a chimney and closet, the room being richly gilded and painted and the outside shutters very strongly barred'.

Critics of the lighthouse pointed out that it looked like something from a Chinese mausoleum. Winstanley retorted that he was confident enough of his light to believe it would withstand 'the greatest storm that ever was'. He should not have tempted fate. During the Great Storm of 26 November 1703, over 17,000 mature trees were felled, buildings were stripped to skeletons, and the Queen was forced to take refuge in a cellar at St James's Palace. Winstanley had set out from Plymouth on an inspection tour of the light and was on the Eddystone when the storm began. According to observers, the light was visible until midnight, when the mountainous seas and spray obscured it from view. The next morning, the only thing that remained of the tower were the twelve iron foundation bars. Winstanley himself had drowned.

Winstanley's successors were more pragmatic. John Rudyerd, who came forward in 1706, was a silk merchant who, like Winstanley, had no prior engineering experience. The design he proposed was for a wooden tower, which, he reasoned, would be flexible enough to withstand the force of the waves. His plans show a conical structure like the base of a windmill on which was perched a simple lantern and an umbrella-like cupola. Unlike Winstanley, Rudyerd concentrated hard on anchoring the tower to the rock, dug deep into the gneiss to lay iron support pillars, and weighed down the whole structure with a rubble ballast. The inner walls of the tower were built with alternate layers of granite and timber, and the outer were sheathed with wooden planks. It was a simpler, more practical design that Winstanley's, and survived for forty-six years despite the constant underminings of woodworm. Unfortunately, it was also flammable. In December 1755, Henry Hall, the oldest keeper on the light, discovered the lantern room on fire. By the time he had roused the other two keepers, the fire had spread downwards through the tower, and the men were forced to scramble out to the slippery shelter of the rock. Eight hours later, they were rescued by a local boatman who had seen the light blazing from the shore. All that was left of the tower was the stump of the base and the metal rods that Rudyerd had used as foundations. Hall, meanwhile, complained urgently that his insides were burning, though his two companions ignored him, believing that the shock of the fire had affected his mind. Twelve days later he died. The postmortem examination discovered a lump of lead weighing seven ounces in his stomach. Hall had, apparently, been staring upwards at the fire in the cupola when a molten ball of lead had fallen into his open mouth.

Almost immediately, the owners of the Eddystone began looking for a successor to Rudyerd. This time, they chose an engineer. John Smeaton was a Yorkshireman born in 1724, and originally trained as an instrument maker. The amateur habits of engineering at the time suited him and after his involvement with the Eddystone he went on to become involved with an exceptional span of projects, including Norfolk fen drainage, the construction of several bridges and aqueducts and the Forth–Clyde Canal. 'I consider myself,' he wrote later, 'in no other light than as a private artist who works for hire, for those who are pleased to employ me, and those whom I can conveniently and consistently serve.' The Eddystone proved his most flamboyant epitaph, and provided the template for all future attempts at lighthouse engineering. Smeaton's plans for the new lighthouse differed again from Rudyerd's. It was essential, he considered, to build the whole structure from stone to ensure that it was heavy enough to resist the worst of storms. Instead of making the tower conical, he designed a tapering base to give it the greatest possible solidity on the rock and the maximum buttressing above. He used the analogy of a tree trunk: 'the English oak tree withstands the most violent weather conditions; so I visualise a new tower shaped like an oak. Why? Because the oak tree resists similar elemental pressures to those which wrecked the [Winstanley] lighthouse; an oak tree is broad at its base, curves inwards at its waist, becomes narrower towards the top. We seldom hear of a mature oak being uprooted.' Its construction, he reasoned, would combine the fit of a jigsaw puzzle with the elegance of a tree trunk's rings. By cutting each block of stone so it slipped snugly into the next, it would be possible to create an almost unbreakable structure, so neatly assembled that the sea would be unable to find its usual destructive leverage. The only flaw in this dovetailed arrangement was that each layer (or course) of close-fitted blocks had to be connected to the next with small oak pins (trenails.) But otherwise, the basic principle of lighthouse engineering was now intact.

The first four years of construction work were dogged with difficulties. At one point Smeaton, returning from the rock, was blown off course by a storm and found himself heading briskly towards the coast of France. But work continued almost all year round. The first course of stone was laid by 1757 and during the winter months Smeaton attended to the cutting and dressing of the rock and the vetting of the workmen he employed. He had a fanaticism for detail and, to the astonishment of the builders, often flung off his jacket and bowler hat and pitched in with the work himself. By October 1759, the lighthouse was complete. It had taken three years, 1,493 blocks of stone, 1,800 oak trenails and £40,000 to achieve; it weighed 1,000 tons and stood 80 feet tall. In the lantern at the top was the light source – 24 tallow candles. One of these is still on display at Trinity House in London; it is roughly the size of the average dinner-table candle.

Fifty-two years later, on his second tour of the English lights, Robert Stevenson reached the Eddystone light. It was, he found, a disappointment. Throughout his time as an engineer, he had worked with Smeaton and the Eddystone in mind. He had read Smeaton's *Narrative of the Construction of the Eddystone Lighthouse*, taken issue with Smeaton's tree trunk and perfected Smeaton's dovetails. He had made comparisons, drawn analogies and spent much time in discord or agreement with his paper hero. It was perhaps unsurprising that the reality did not live up to imagination. Once landed, he noted almost peevishly that 'the appearance of the Eddystone is rather diminutive.' On the exterior, 'the faces of the stones are in some instances wasting and the joints are extremely coarse and not neatly wrought,' and upstairs, 'the doors and windows are ill-hung and ill-fitted and the iron work much rusted.' Five years later, he returned again. This time, though the tower seemed in order, 'in the interior everything remains as when the tower was built. The stair is inconveniently narrow, the interior is too small and the whole furniture and apparatus is coarse, paltry and mean.' This, he concluded, was not Smeaton's fault, but that of Trinity House, who had now bought back the light and were responsible for its maintenance. Robert also made a thorough inspection of the reef itself.

> It is shaken all through, and dips at a considerable angle, perhaps 1 in 3, towards the south-west; and being undermined on the north-east side for several feet, it has rather an alarming appearance ... There is a hollowing of the rock which penetrates at least to the circumference of the base of the lighthouse ... The rock here projects beyond the base above and not below, which must give the sea a great hold. I therefore conclude that when the sea runs high there is danger of this house being upset after a lapse of time when the sea and shingle have wrought away the rock to a greater extent ... Were I connected with the charge of this highly important building, I must confess I should not feel very easy in my mind for its safety.

Robert was, as usual, right. Smeaton's work on the lighthouse was impeccable; this time it was the rock itself that was at fault. The battering of the waves had not only worn the reef but jolted the tower, which now visibly shook with each new storm. Trinity House inspected it and twice strengthened the cavity that Robert had found. It made no difference. In 1878, James Douglass, Trinity House's Chief Engineer, announced that the tower had deteriorated to such an extent that it would have to be rebuilt again. The completed tower stood 168 feet above high water, used 2,171 blocks of granite weighing 4,668 tons and cost £59,250. It was almost twice the height of Smeaton's tower, and incorporated nine rooms inside. Finally on 18 May 1882, the last of the Eddystone lights was lit. After five designs, four architects and two centuries, the Eddystone passed into the land of legends.

Robert's fascination with the Eddystone was not just professional curiosity. Even in 1801, he had a project in mind that was increasingly preoccupying him. His studies of Smeaton's methods showed that what had been done in England could also be done in Scotland. He was also confident enough even then to believe that Smeaton could be bettered. The Eddystone proved that lighthouses could be built in apparently impossible conditions, could be constructed to be both tough and flexible, and could withstand the sea for hundreds of years. It only needed an inspired designer to bring the Eddystone's example to perfection. And Robert, quietly at first and then with gathering conviction, knew he was that man.

FOUR

The Bell Rock

One day in January 1793, a letter arrived with the Commissioners of Northern Lights. Captain Alexander Cochrane of the Royal Navy was in charge of His Majesty's ship *Hind*, one of two major warships presently stationed on the east coast. As a military man, he was not given to over-excitement, but his letter sounded a compelling note of alarm. 'I think it a duty I owe the public,' he wrote, 'to call your attention, as Trustees for the Northern Lights, to the great hazard and peril that the trade of the East of Scotland is subject to for want of a lighthouse being erected on the Bell or Cape Rock.' He had himself made investigations into the area and thought carefully about the expense of building a light. Cochrane concluded that the responsibility for muzzling the Bell Rock lay squarely with the Commissioners. It was, he added ominously, a significant obstacle to good trade and proper communications with Scotland at present, and the Commissioners should remain mindful of their duties as free-market men.

Cochrane was not alone. Six year later, one of his colleagues, Captain Joseph Brodie, became so concerned by the vicious reputation of the Bell Rock and the inaction of the Lighthouse Commissioners that he decided to take matters into his own hands. Having prepared a model of a light which was to rest on four cast-iron pillars and be bound together with metal strapping, he presented it with due ceremony to the Commissioners. The Commissioners rejected it. Undaunted, Brodie ignored them and sailed off to the rock where he built two temporary wooden beacons. Almost immediately, the sea destroyed them. Brodie then enlisted the support of several Leith shipowners, and sailed out again. This time, he prepared a sturdier framework made of timber and iron and planted well into the rock. The beacon lasted for five months before being sunk by a hard winter. Brodie, still unwilling to admit failure, returned to the Commissioners and offered to build yet another light, this time constructed entirely out of iron. The Commissioners turned the offer down, but promised to reimburse him a small sum in recognition of his efforts, however apparently unproductive. Brodie presented a bill for several thousand pounds. The Commissioners, outraged by the demand, told him he could take £400 or nothing. Brodie refused it. Eventually, after years of futile dispute, the £400 was paid to his widow. And still no light appeared on the Bell Rock.

Cochrane's complaints and Brodie's thwarted efforts had not been lone protests. Both were accompanied by a flurry of petitions from like-minded men. Captains, shipowners, sheriffs and landowners had all written to the Commissioners at various times pleading in tones of mixed desperation and severity for a light on the Bell Rock. Ever since the Board's establishment in 1786, petitions had been raised and schemes debated, running through the Minute books in a plaintive refrain. Endless solutions were proposed: replacement bells, lights on stone or wooden pillars, even lights on rafts. The suggested designs varied from simple beacons to elaborate contraptions built to impossible specifications. The reason for their agitation was obvious – the Bell Rock wrecked ships, year after year; it was a danger and a discouragement to shipping in the area. Sooner or later, the Commissioners were going to have to do something about it.

Both the shipowners' alarm and the Commissioners' intransigence were understandable. The Bell Rock, a sharp sandstone reef twenty-seven miles east of Dundee and eleven miles south of Arbroath, was indeed deadly. In all, the rock extended for around 1,400 feet and was shaped roughly like a slice of cheese turned on its side. Its danger lay in the fact that for much of the time, the reef was underwater. Even at low tide, its rocky spikes protruded only a little way out of the sea, and the whole area was only visible from a distance because of the ill-tempered currents and white water lacing its surface. At high tide, the reef vanished under seven foot of water, leaving nothing except the occasional flash of spray to betray its presence. Even in the mildest weathers, the submerged sides set up a whirligig of eddies into which unwary ships could be drawn. By the turn of the century, the Bell Rock's habit of demolishing ships coming in or out of the Firth of Tay had given it a notoriety far beyond Scotland. As Louis later explained, the rock was positioned unusually awkwardly. 'Placed right in the fairway of two naviga-tions, and one of these the entrance to the only harbour of refuge between the Downs and the Moray Firth, it breathed abroad along the whole coast an atmosphere of terror and per-plexity; and no ship sailed that part of the North Sea at night, but what the ears of those on board would be strained to catch the roaring of the seas on the Bell Rock.' Those who tried to avoid it often sailed too close inland and grounded themselves on the jagged coastline nearby. Those who sailed too far to the west gave up the chance of refuge from the North Sea. Unsurprisingly, it had become a favourite among the local wrecking population, who, providing they could escape being wrecked on the rock themselves, found it a fruitful source of loot. On average, it was calculated, the Bell Rock ruined up to six ships every winter.

Over the course of the centuries, various attempts had been made to mark the Inchcape Rock, as it was originally known. During the fourteenth century, the Abbot of Aberbrothock (Arbroath) fixed a bell on the reef to warn passing sailors. A Dutch pirate, intent on accumulating as much wreck as possible, removed it. 'A yeare thereafter,' an early Scots historian noted smugly, 'he perished upon the same rocke with ships and goodes, in the righteous judgement of God.' The Inchcape was gradually retitled in honour of the bishop, becoming instead the Bell Rock. The poet Robert Southey, impressed with the legend of its naming, wrote *The Ballad of the Inchcape Rock* in 1815 after a visit to Scotland.

> Sir Ralph the Rover tore his hair
> He curst himself in his despair,
> The waves rush in on every side;
> But the ship sinks fast beneathe the tide
> For even in his dying fear
> The dreadful sound could the Rover hear
> A sound as if with the Inchcape Bell
> The Devil below was ringing his knell.

Though perhaps not his finest poetic moment, Southey's contribution only added to the Inchcape's myths.

Reality also intervened. In December 1799, a gale lasting three days destroyed over seventy ships around the Scottish coast. The warship HMS *York*, caught offguard by the storm, ran aground on the Bell Rock and sank with the loss of all on board. The resulting furore in Parliament and the rising pressure from local shipowners and merchants made the Northern Lighthouse Commissioners increasingly uncomfortable. But even the loss of a major warship did not appear to move them. Over the years, they had become so accustomed to the rumble of complaints that they had a standard reply prepared. They were, they said, most attentive to the concerns of seafaring men, 'but their funds were never in a situation to attempt so expensive and hazardous an enterprise'. At present, they had scarcely enough money to pay for a candle, let alone something as complex and sophisticated as a lighthouse eleven miles out to sea. All their funding came from dues levied on shipping, and those barely scraped the costs of maintaining the existing lights. In addition, they needed the permission of Parliament for any further lights, which meant drafting and presenting a special bill and then waiting for months, maybe years, while it made its slow passage through the debating chambers. Besides, they added privately, the Bell Rock was an impossible site, there wasn't the engineer in existence foolhardy enough to attempt even a platform on it, and that was an end of it. Anyone who had actually seen the place would realise it was an absurd request. The Bell Rock was deadly, and that was that.

Their chief engineer, however, had other ideas. Robert Stevenson believed that a light on the rock was both possible and necessary. From his long experience with the Northern Lights, he argued, he was eminently qualified to decide on exactly where and how to build lighthouses. A tower on a sub-merged rock was an awkward endeavour, but not an impossible one. His recent visit to the English lights had only confirmed his view. If Smeaton had managed to build a lighthouse on the Eddystone then he, Robert Stevenson, could go one step further and build one on the Bell Rock. It would indeed be expensive, but the Commissioners should weigh the cost of a lighthouse against the price of another century of wrecks and deaths. Furthermore, as he was fond of pointing out, he considered himself the best qualified candidate to build the light. He had been applying himself to the science of lighthouses now for over ten years; he understood them better than any other engineer in Britain, and he had absolute faith in his own judgement. Once confronted with a true test of his abilities, Robert rose to the occasion.

From 1800 onwards he lobbied, cajoled and pressured the Commissioners for the chance to design and build a suitable lighthouse. The NLB meetings, usually staid affairs devoted to finance and keepers' grazing rights, suddenly filled with the chatter of experiments, addresses and arguments over the Bell Rock project. With his customary mixture of bluntness and guile, Robert appealed to anything and everything that he felt might possibly strike a chord with the Commissioners; public duty, maritime prosperity, benevolent humanity, compassion or guilt. In a lengthy address presented to the Commissioners in 1800, Robert deployed all his political skills towards his cause. A good dose of melodrama, which he usually avoided, also helped. Mariners, he announced, often avoided the terrors of the Firth of Forth 'on account of the Bell Rock, which, like another Cerberus, guards its entrance'. 'I will venture to say,' he continued, 'that there is not a more dangerous situation upon the whole coasts of the Kingdom, or one that calls more loudly for something to be done, than the Cape or Bell Rock.'

Fine words were only half the battle. As he pointed out in the report, he had also studied the rock carefully in his own time, and had concluded that a lighthouse could be constructed on at least one part of the reef. In 1794, he had tried to reach it, but, to his embarrassment, had been unable to land because of the filthy seas. Even from a distance, however, Robert had concluded that it would be possible to use the reef's own geography to his advantage. His first idea was for a lighthouse on pillars, similar to the model of the Smalls light in England. Six pillars of cast iron would be sunk into the sandstone, supporting a small, rocket-shaped cabin with a light-room on top and four small apartments for the keepers. The advantage to a pillared light, Robert considered, was that it would provide least resistance to the impact of the sea; waves would rush through the pillars, instead of banging heavily against a solid structure. The flaws to this scheme, as he acknowledged, were that ships might become entangled within the pillars and that the cast-iron foundation would rust, needing frequent maintenance. The Commissioners, considering the plan both too flimsy and too expensive, rejected it.

Robert's second trip to the reef in 1800 was more successful. He was escorted by local fishermen, who managed to land him on the rock and then disappeared to hunt for wreck while Robert made his investigations. By the time Robert had concluded his survey, he noted, the fishermen had gathered 'two cwt of old metal, consisting of such things as are used on shipboard,' including, 'a ship's marking iron, a piece of a ship's caboose, a soldiers bayonet, a cannon ball, several pieces of money, a shoe buckle . . . a kedge anchor, cabin-stove crowbars &c.' If nothing else, their findings impressed on him the rock's ability to destroy even the sturdiest of ships. Robert's findings were equally fruitful. He had surveyed the rock as fully as possible, noting at least one site on which a building could stand. He also realised that the durability of the sandstone made a perfect base for the light. Back at home in Baxter's Place, he abandoned his plans for a pillared light and turned instead to the idea of a stone tower. It would be built in much the same way as Smeaton's Eddystone light, using dovetailed granite and tapering gently upwards towards the lantern. It was, he conceded, to be 'a work which cannot be reduced to the common maxims of the arts and which in some measure stands unconnected to any other branch of business'. The plan he presented to the Commissioners in his report differed from Smeaton only in details. He would diminish the width of the walls gradually, instead of following the Eddystone's pattern of a solid stone base with the apartments perched on top. A stone tower, he concluded, was better able to resist the pressures of the Bell Rock, and would last longer than any structure of metal. 'The more I see of this Rock,' he confided privately to John Gray, clerk at the NLB, 'the less I think of the difficulty I at first conceived of erecting a building of stone upon it.'

His report to the Commissioners conceded that the costs would indeed be high – his initial estimate came to £42,685 – but that it would be counterproductive to try to cut costs in such a 'matter of importance to the whole mercantile interest upon the east coast of Great Britain'. He also spoke to the Commissioners in language calculated to move them. 'The most material general purpose, however, which would be answered by this light is the opening of the Firth as a place of safety . . . As Commerce and Industry are considered essential to the wealth and happiness of the community, every effort to assist the Mariner, in his course through the pathless ocean, will be regarded both as the call of interest and humanity.' Just for good measure, he reminded the Commissioners of the loss of HMS *York*, 'this fatal catastrophe, of which history affords few examples, is the more to be lamented when it is considered that a light upon this rock would infallibly be the method of preventing such dreadful calamities . . . Until this blessing comes though your hands as Commissioners for erecting light houses in the Northern parts of Great Britain, it is much to be feared that the cause continuing, the sad effects will not cease.' Finally, Robert concluded humbly that 'in offering you this address, I act from considerations of duty . . . I am ready, in conjunction with Mr Thomas Smith, if called upon, to come forward with my services in the way you may judge most advisable.'

Though a masterpiece of Robert's lobbying skills, the address left the Commissioners unmoved. The expense was still too great, they argued, the enterprise too dangerous and the precedents insufficiently tested. Furthermore, they added crushingly, even if the project was to go ahead, it would not automatically go to Robert. They suggested that they would 'apply to the most eminent of their profession, as well mariners as civil engineers, for their opinion as to what manner and with what materials it would be most proper to erect the lighthouse'. As they pointed out, Robert had only built one lighthouse on his own and, despite his decade's experience in design, planning and management of the Northern Lights, was still a young and relatively unqualified engineer. Although he was the NLB's chief engineer, Robert was only on contract to the Board, and had no God-given right to suppose he would be allowed to undertake the work. The Commissioners, although no engineering experts themselves, found Robert's bumptious self-belief disconcerting. As they reminded him, he was only a small pawn in a much grander game, and the Commissioners were cautious men.

Nevertheless, Robert's report had made the Commissioners thoughtful. Despite their wariness, they remained conscious of their public duty. It was not only Robert, after all, who was prodding hard for the project to go ahead; the queue of vested interests grew louder by the month. By 1803, the Commissioners unbent enough to allow a Bill to be presented in Parliament, only to have it thrown out by the House of Lords. They also sought the advice of other, more reputable engineers. Thomas Telford, the famed architect of Highland roads and bridges, supplied a sketchy estimate for a stone lighthouse costing £29,000, and then disappeared, pleading overwork. John Rennie, the Commissioner's next choice, came and stayed.

By 1805, Rennie was forty-four, a shrewd and charming man born and raised in East Lothian. He had trained first as a millwright, and then taken work as a jobbing engineer, working on bridges, canals and steam power. He had designed and built the Crinan Canal in Argyll and been jointly responsible for widening the Clyde to allow for deeper-hulled vessels to reach the Glasgow ports. By 1800 he was being courted for projects throughout Britain, and spent his time in a flurry of travel and correspondence. Samuel Smiles, in his *Lives of the Engineers*, records Rennie's painful devotion to his profession. 'Work was with him not only a pleasure,' he wrote, 'it was almost a passion. He sometimes made business appointments at as early an hour as five in the morning and would continue incessantly occupied until late at night.' In thirty years, Rennie took one holiday, a trip to France during which he spent most of his time gathering information on Napoleon's docks and harbour works. He, like Robert, had learned his trade through the practice of it; he, like Robert, had the same belief in experiment and discovery which characterised engineering at the time. Like Robert, he had also come up against the intransigence of those in power, and was well used to scepticism and obstruction. Smiles recounts the comment of one of Rennie's friends. 'What I liked about Rennie,' said the man, 'was his severe truthfulness.' The Commissioners of Northern Lights, scanning the narrow lists of well-known engineers, found in Rennie a perfect solution. He had, as yet, no experience of building lighthouses, but his skill with bridges and marine works and his ability to turn his mind to anything were recommendation enough.

Robert, meanwhile, was torn between wariness, delight and piqued ambition. Rennie's appearance was evidently dispelling the last doubts of the Commissioners. And yet Rennie's confidence seemed to have cancelled Robert's hopes of building the light. He knew he would be involved in the works in some capacity, but as the Commissioners' enthusiasm grew, his own faltered. A friend who wrote to congratulate him on the progress of the project received a jaded reply. 'Your congratulations relative to the Bell Rock business I am sensible of, but . . . I have certainly bestirred myself in this business for eight or nine years. Yet now, when the matter seems at hand, I look forward with much anxiety to the personal dangers and numberless difficulties which must be struggled with by all engaged in this work.' Robert battled for months with his competing demons; the knowledge that he was himself only the paid employee of the Commissioners, the belief that Rennie was indeed the most suitable man for the job, and his own unshakeable certainty that he alone understood the subtleties of the Bell Rock. He was a professional man, and should conduct himself as such, but he also nursed a fierce ambition for his dreams.

In December 1805, Robert wrote almost sorrowfully to Rennie, enclosing a copy of his plans and remarking that the Bell Rock was 'a subject which has cost me much, very much, trouble and consideration . . . should you wish any explanation or further enquiries in which you can employ me, everything shall be allowed to stand in order to further your views.' At the same time, he began a little diplomacy with the Commissioners. As he wrote to Alex Cunningham, Secretary to the NLB, he had been in touch with Rennie 'stating that it was thought he would consider this as a situation where the personal danger was so great and the difficulties to be struggled with so many as to be entitled to at least double what is paid for Sea works on the Shore which I think was exactly your ideas upon a subject which much interests my Family and future prospects, I may say, in life.' The Commissioners seemed unimpressed with the possible dangers to Rennie's health and stayed silent. From time to time, Robert presented a further plan or report, urging himself on the Commissioners' attention as often as possible. The Commissioners noted his contributions and then ignored them. But Robert had no intention of giving up.

Initially Rennie was asked to provide a survey and report, and to give his recommendations on the most suitable structure for the reef. Despite being thoroughly seasick on his way to the rock, he, like Robert, concluded that it would be practical to build a substantial structure on the Bell Rock. 'I have,' he wrote in his report to the Commissioners, 'no hesitation in giving a decided opinion in favour of a stone lighthouse.' He also estimated that it would cost around £41,840, only a few hundred pounds less than Robert had calculated, and dismissed Telford's brief assessment as farcically inaccurate. 'He has,' he wrote to Robert, 'no originality of thoughts, and has all his life built the little fame he has acquired upon the talent of others, which he has generally assumed as his own.' Rennie's plans for the light followed Robert's conclusions in most essential respects, though he stuck much closer to the template set down by Smeaton on the Eddystone, 'its general construction, in my opinion, rendering it as strong as can be conceived'. Robert's plans, by contrast, seemed less cautious than Rennie's. The longer he spent studying the site, the bolder he had grown. Rennie's involvement, however, appeared to remove all Robert's confidence in one neat blow.

But Rennie's verdict on the feasibility of the light had shifted matters dramatically. The Commissioners, faced with a definitive second opinion, succumbed. They began preparations for a second bill, ensured it had sponsorship and then waited while it made its languid passage through the House of Commons. Objections were raised by those who thought that public money would be better spent on Britain's defences against Napoleon. Both Rennie and Robert were sent south by the Commissioners to answer for the project, and to lobby for the maximum funds. Four months later, in July 1806, the bill authorising construction was passed and the Commissioners permitted to borrow £25,000. Finally, the ultimate sanction arrived. In the General Meeting on 3 December 1806, the Commissioners announced 'That the building to be erected for the purpose of a light house on the Bell or Cape Rock shall be of Stone, and that the work shall be vested under the direction of John Rennie Esq, Civil Engineer, whom they hereby appoint Chief Engineer for conducting the work.' Mr Stevenson, they added in a cursory postscript, 'was authorised to proceed along with Mr Rennie, and to endeavour to procure a yard and the necessary accommodation at Arbroath.'

During those long months of waiting, Robert had continued to work on his plans for the light. He had been much affected by Rennie's openness towards him, and the two had tried to be generous in their dealings with each other. Rennie was, after all, a kind-hearted man; after the untimely death of another of Robert's children, Rennie had written to him offering support and consolation. 'I am truly sorry for the affliction you have lately sustained in your family, an affliction as a Parent I well know ... But, my good Sir, this is a world of trial, and when we consider our Situation we ought not to mourn or repine at the dispensations of Providence; we are here only in a state of probation, to prepare for another and a better world.' But whatever rapport there was between the two in private could make no difference to Robert's professional quarrel with Rennie. Kind words could never compensate him for the frustration of his hopes.

Though sorely wounded by the Commissioners' indifference to his efforts, Robert took the insult quietly, preferring instead to scheme alone. He now began adapting and refining his designs even further, making allowances for the different conditions on the rock and his own sophisticated knowledge of local materials. As it became evident that the Commissioners intended using him as a kind of stage-manager for the works – ordering materials, hiring men, dealing with suppliers – Robert also discovered a new form of leverage against his chief engineer. He sent his plans to Rennie, along with an exhaustive list of questions and demands: what kind of stone to use, which quarry was it to come from, how many men would be needed, which measurements, which type of mortar. Rennie replied with corrections and further suggestions. Robert, aware that Rennie was now distracted with other matters, ignored his corrections and bombarded him with an urgent list of further questions. Rennie wrote back again, less promptly.

By 26 December – just over three weeks since the original meeting – the two were already appearing jointly in the Minutes. 'Mr Rennie,' the Clerk noted, 'proposed to the meeting that Mr Stevenson should be appointed assistant engineer to execute the work under his superintendence.' The Commissioners, making allowance for Rennie's other projects, agreed. Seizing the advantage, Robert wrote again with a ceaseless series of letters, reports, memorandums and queries. He needed Rennie's decision, he wrote, on the quality and quantity of stone for the light, measurements for timber, his opinions on various types of Aberdeen granite, his estimate for the workmen's accommodation, an inventory of tools. He needed measurements, calculations, assessments, and he needed them immediately; further delay would only hold up the works for yet another winter. His tone was polite but plaintive; most of his correspondence carried an undertone of complaint. Rennie, he implied, was frittering his time away on butterfly-minded schemes while he, Robert, remained behind to do the real work.

Rennie, away from Edinburgh and hopping frantically from one project to another, replied when he had the chance. He had been to visit another offshore light in Ireland, he wrote in a letter of September 1805, noting warningly that 'the Bell Rock will be ten times as hard to do as the lighthouse on the South Rock can have been'. When he did have the chance to reply, he had evidently thought deeply about the different methods of constructing the light, and of the best use of materials. His letters to Robert were filled with an answering barrage of estimates, queries and opinions; whether northern granite was really the most appropriate stone, how the lantern was to be designed, how best to mix mortar, how many link-pins should be necessary. Robert, in his turn, proposed a series of changes and amendments to Rennie's schemes and blithely ignored many of his chief engineer's suggestions. Anyone following the proceedings closely might have noticed that it was not Rennie's plans that Robert was working to by now, but his own. Rennie, startled by the bombardment of correspondence and by Robert's peremptory tones, became distracted and uncertain. As the date for the start of works on the rock neared Robert increased the pressure further. Rennie, away in London, Bristol, Portsmouth, Glasgow or Birmingham, kept up as best he could. Though resolutely good-natured throughout, he clearly began to feel his grip slipping as Robert's tightened.

Nor did Robert neglect the Commissioners. By March 1808, while works were already well underway, Robert felt himself to be in a strong enough position to try one further manoeuvre. In an address to the assembled lighthouse trustees, Robert suggested – in his usual circumlocutory manner – that he should resign his other business and be appointed sole engineer to the Northern Lights. The lighthouse work, he argued, had expanded so greatly since Thomas Smith was first appointed that it no longer suited anyone to have a mere contracted amateur supervising all aspects of the lighthouses. The Commissioners needed someone to devote themselves full-time to the position; Robert, naturally, was the rightful candidate. And, while they were about it, perhaps they could consider the question of his salary? The move was a shrewd one. Not only did it consolidate Robert's position, but it also removed the Commissioners' belief that lighthouse work could be offered to any old jobbing engineer. True, it was too late for the Bell Rock, but still, the Commissioners could no longer reduce Robert to a footnote in their dealings.

The whole episode, from Rennie's first appointment to the final realisation of Robert's ambition to be sole master of the Bell Rock, took three years – three years of agile diplomacy and shrewd manoeuvrings; three years of quiet stubbornness. In public, Robert remained self-effacing, pragmatic in his dealings and stoic in his workload. But his sly campaign to sideline Rennie and ensure that he would be credited with sole authority for the Bell Rock demonstrates a flintier habit of mind. He was competitive, ambitious and frequently political; he knew the value of his own work and expected its due recognition. He planned many moves in advance, and he had the self-assurance to believe – before the first surveys of the site had even been completed – that the Bell Rock would turn out to be his passport to immortality. The fact that he was to be proved right makes him admirable; it does not always make him likeable.

Having effectively removed Rennie and dispensed with the distractions of his other engineering business, Robert focused wholly on his pet project. After deciding that a tower of stone was possible, he set about establishing the neatest methods of organising the works. All building work was circumscribed by time and tide. Since it was impossible to get near the rock in the winter months, Robert aimed to do as much as he could between May and September of each year. His plan, to which Rennie had vainly objected, was to build temporary workmen's quarters next to the tower. This, he argued, would allow the builders to stay for longer at the site and short-circuit the need to make endless shore trips for supplies or rest. The Commissioners clearly thought the idea of a workmen's barracks was the lunatic fantasy of a deluded mind, but, with a little more argument, Robert was given the authority to start. They were more amenable to the construction of a makeshift beacon near the rocks to provide early warning for sailors, since it would allow them to charge shipowners for its use and therefore provide an extra source of funds.

By now, Robert was working openly from his own plans. With a little sleight-of-hand, Robert had managed to convince the Commissioners that his designs were more trustworthy than those of Rennie; the Commissioners, distracted by other business and wary of the prospect of another of Robert's sonorous fifty-page reports, capitulated. They did, however, express serious reservations at Robert's digressions from the Eddystone template. Robert argued that the comparison between the Eddystone and the Bell Rock was tempting but inaccurate; the Bell Rock was almost permanently submerged underwater, it had to withstand fiercer seas and Smeaton's technique of dovetailing the stones was, albeit only slightly, flawed. Robert's light would be taller and more sturdily built. The joins between the different floors would be stronger, and the whole tower more substantial. The Commissioners complained, and then, under pressure, let the subject go.

For two years, Robert occupied himself with men and materials. Over a hundred men, many of whom Robert had chosen from previous lighthouse projects, had to be appointed and trained, and many had their own misgivings. Two of the sailors that Robert had been considering heard that they would be working on the Bell Rock and were so terrified by its fearsome reputation that they fled and never returned. As the Commissioners had instructed, a workyard was established at Arbroath and a lighthouse vessel, christened the *Smeaton*, commissioned to take the builders, joiners, mortar-men and smiths to and from the rock. Robert took an interest in every part of the plans, fixing itineraries, considering workers, examining stones, drawing and redrawing his plans. An early inventory is revealing, both as proof of Robert's meticulousness and for the sparse equipment used on such an exceptional scheme. The workmen's tools included 44 pick-axes, 11 stone axes, 9 boring hammers, a smith's forge, 2 vices, a horse and harness and 236 balls of lead. Robert also noted precisely the five screwdrivers and around 400 different types of nails.

Once he had obtained authority for the start of works, Robert's letters and orders became even more urgent. Workers, foremen and suppliers were assailed by a daily flurry of demands. Why, he wrote, did they not move faster? Why had he not received a speedier reply to his last letter? Why did it take so long to quarry the stone? Why could the boat not be launched? Why did the weather stay so troublesome? Why was the sea so uncooperative? His writing slipped deeper into illegibility and his grammar, never strong in a crisis, collapsed entirely under the strain. Those who did reply in time found themselves submerged under a further blizzard of correspondence. Whatever his staff did, it appeared almost impossible to satisfy Robert. Impatience remained one of his most troublesome characteristics, and the one which he struggled hardest to contain. His occasional dissatisfaction with humanity made him, at times, a difficult employer. The Bell Rock works, unsurprisingly, therefore become an exacting workplace; the men could be assured of fair treatment, but were also expected to work long hours in foul conditions for a hard-minded man.

One example of this was the rations permitted to each man. The list Robert drew up was based on naval diets at the time, which had often been criticised for their parsimony and lack of nourishment. Even so, sailors on His Majesty's warships were allowed a full daily allowance of bread, pork and cheese, plus a gallon of beer each evening. At the Bell Rock, by contrast, each man was to be allowed a meagre ration of ship's biscuit, beef, oatmeal, barley, and a few vegetables; once a week, they were allowed a little beer. Similarly, work on the reef began at 5 a.m. in mid-summer, and would often end after 8 or 9 p.m. if the weather was fine. Even Robert's stern belief that the Sabbath should be kept sacred was subsumed to the pressure of work.

Sundays were to be working days, he dictated, the same as any other. 'Surely,' as Robert remarked later by way of justification, 'if under any circumstances it is allowable to go about the ordinary labours of mankind on Sundays, that of the erection of a lighthouse upon the Bell Rock seems to be one of the most pressing calls which could ... occur, and carries along with it the imperious language of necessity.' Robert's tolerance of the builders' occasional fits of superstition or protest was naturally limited.

The first full day of work, 16 August 1807, was marked with a modest celebratory ceremony. 'The piers, though at a late hour, were perfectly crowded,' wrote Robert in his journal, 'and just as the *Smeaton* cleared the harbour, all on board united in giving three hearty cheers ... The writer felt much satisfaction at the manner of this parting scene; though he must own that the present rejoicing was, on his part, mingled with occasional reflections upon the responsibility of his situation.' The first task was to prepare the rock for the foundations of both the workmen's barracks and the tower. In addition to the dense scribbles of seaweed, the men discovered 'a great variety of articles on the rock, [and] some silver and copper coins. On Tuesday they found a ship's marking iron lettered JAMES.' Once uncluttered, the rock became, if anything, more dangerous, its surface so slippery that even the mildest fall meant injury.

The sea was also troublesome; the workmen became used to seeing a morning's work swept away by a single wave, and to debilitating binges of seasickness. Until the barrack was completed, the workers were forced to stay on the *Smeaton*, which heaved like a funfair ride in bad weather. Having endured twenty-seven hours of one gale-force storm, during which he was hurled to the floor 'in an undressed state' several times, Robert suggested sailing for the shore. The captain informed him cheerily that they would very likely end up suffering the same fate as any other sailor round the Firth of Forth, and either dash themselves to pieces on the Bell Rock itself or on the Isle of May. The gale lost them ten days of work in all, and when Robert finally returned to the rock, he discovered that six vast blocks of granite and the smith's anvil had been dislodged from their places and hurled to opposite ends of the reef. Robert's scepticism about grandiose claims of the sea's power was being replaced with a lively respect.

Even when on the rock, the workmen were not always safe. In one particularly alarming incident, the *Smeaton* came adrift from her moorings while thirty-two of the men were working on the foundations. Robert, who was standing on a ledge a little further off, noticed the boat bobbing away from the rock and realised instantly that, without her, the workmen would almost certainly be drowned as the tide rose. He stood speechless, torn, as he wrote later, 'between hope and despair'. If he warned the men, he knew, they would only panic and try something foolhardy. So the oblivious builders hammered on, and the boat drifted further and further away. Finally, when the tide had risen so far that work became impossible, the men gathered by the moorings for the two smaller boats,, normally used only to transport provisions. As they did so, they realised what had happened. 'Not a word was uttered by anyone, but all appeared to be silently calculating their numbers . . . the workmen looked steadfastly upon the writer and turned occasionally towards the vessel, still far to leeward. All this passed in the most perfect silence, and the melancholy solemnity of the group made an impression never to be effaced from the mind.' Robert was about to shout instructions to the men – to remove all ballast from the remaining boats, to cling to the gunwales – but found himself mute with fear. As he stooped down to moisten his lips with water, 'someone called out "a boat, a boat!" and, on looking around, at no great distance a large boat was seen through the haze making towards the rock.' It was, in fact, the supply boat, which had arrived purely by chance with a consignment of letters. The men piled in, unhappily aware that most, if not all of them, would have been left to drown if they had stayed much longer. Robert remained haunted for the rest of his life by the experience. Though he passed it off calmly enough at the time, it made him doubly aware of his responsibilities and of his powers. James Spink, captain of the supply boat, was rewarded afterwards with a lifelong pension and a full lighthouse uniform. The men, meanwhile, had to be persuaded with some force ever to set foot on the rock again.

For the first month, the men began constructing the temporary beacon and the iron pillars for the workmen's barrack. Once the beacon was lit, work began on the foundations for the tower itself. While the diggers and borers picked away at the unyielding surface of the reef, Robert busied himself with logistics. Landing the stone blocks, he discovered, was often troublesome – in a pitching sea, transferring one-ton lumps of granite from the cargo boat to the rock was both fiddly and dangerous. And, since each block had been individually cut and shaped to slot into its dovetailed groove, any damage was fatal. Robert tried various solutions, including floating the blocks on cork at high tide in the hope that they would settle on the reef at low tide, and sinking a couple of stone-filled containers onto the rock. Neither method was satisfactory, so eventually Robert suggested the construction of a cast-iron railway to ferry blocks from the landing-place to the foundation pit. It was a costly but useful solution, influenced by Robert's fascination with railways. More conventionally, Robert used a horse and cart to draw stones or deliver provisions at the Arbroath workyard. As a result, Robert developed a great soft spot for Bassey the horse and considered his contribution to the works as valuable as any of the men's.

By the time work was abandoned in early October 1807 for the winter, Robert was satisfied by progress. The temporary floating light was now ready and working, the workmen's barracks was nearing completion and work was progressing well on the foundations for the tower. The day before Robert left, John Rennie appeared for a tour of inspection. 'I propose myself much pleasure in the viewing of your operations,' he wrote to Robert shortly before his visit. 'This will be heightened if in the interim you can bargain with old Neptune to favour us with a quiet sea while I am on board the floating light. I hate your Rolling Seas ... however, my good Sir, we are so tossed and tumbled about in the good Theatre Of Life that it must be taken rough and smooth as it comes our way.' He added a brief, awkward postscript, commending Robert and the Bell Rock to their ghostly mentor, Smeaton: 'Poor old fellow, I hope he will now and then take a peep of us and inspire you with fortitude and courage to brave all difficulties and dangers, to accomplish a work which will, if successful, immortalise you in the annals of fame.' He spent a night in strained bonhomie with Robert on the lighthouse yacht; Robert later mentioned that he 'enjoyed much of Mr Rennie's interesting conversation'. It was to be one of only three visits that Rennie made to the Rock during the four years of its construction.

Throughout the winter, Robert busied himself with other lighthouse business, and with his family. Thomas Smith was now living out his retirement at Baxter's Place, preoccupied with his recreational but time-consuming New Town pursuits. For Robert, as head of the Smith-Stevenson brood, Baxter's Place was still a satisfying home to return to. Most of his business was conducted from the house, and once the flocks of children had been tidied away, he used it as both office and headquarters. There were also many people keen to meet him. As the reputation of the Bell Rock grew, so Robert's stature in Edinburgh society increased. Despite his temporary resignation from the Smith & Stevenson lamp-making business, offers of work came in as fast as ever. Robert turned down most projects, but still found time to take an interest in developments elsewhere and to advise the occasional petitioning official.

Robert's campaign against Rennie had almost certainly been conducted with the aim of raising his public profile. He was beady enough to know that the sea-swept reef eleven miles off the coast would make or break his reputation. If his gamble succeeded, it would transform him from a mere freelance engineer into one of the elite new breed of technological experts. And, with Rennie sulking from a safe distance, it appeared as if Robert's fondest predictions were all coming true.

The Lighthouse Commissioners, meanwhile, kept Robert as preoccupied as always. Though they expected him to work full-tilt on the Bell Rock, they also saw no reason for him to give up his duties with the other lights. During the winter, he managed a hurried tour round the coast, checking over the details of each lighthouse. Some of the keepers complained of lack of adequate pasturage for their cows, others of the difficulties in getting schooling for their children or work for their wives. Some had to be reprimanded for incompetence, while new keepers had to be interviewed and appointed. Robert, as usual, also checked sites for new lights and began negotiations with local landowners for buying lighthouse land. Once back in Edinburgh, Robert was expected to justify his Bell Rock expenses and to maintain regular visits to the site throughout the winter. He was also preoccupied with the distractions of the press-gangs, who had discovered the works at Arbroath to be a fertile new source of recruits. Though Robert's staff were theoretically exempt from impressment, it invariably took a lengthy session in the Admiralty Court to extract them; one workman spent five months in prison before Robert could rescue him.

At the end of March 1808, Robert made a thorough inspection of the rock and was satisfied to note that none of the previous season's building work had been dislodged through the heavy winter gales. 'This,' considered Robert, 'was a matter of no small importance to the future success of the work.' It meant both that his calculations and gambles had paid off, and that a solid structure, when correctly designed, could survive the worst that the reef and the sea could hurl at it. Fortified by this knowledge, Robert began full-time work again on the rock at the end of May. The beacon, with its six clawed legs, rose steadily heavenward. In addition to the temporary light, it also now housed a platform on which the smith could work his forge, and a space for the mixing of mortar. Above the platform was to be the workmen's barracks, with four small rooms for eating, sleeping and washing.

This season, Robert concentrated mainly on completing the foundation-pit. The huge circular hole in the reef was comparatively shallow (around two feet deep), but the diggers still had to contend with uncharitable sandstone and contrary seas. As much time was spent baling the pit out as it was in working it; one high wave or an approaching tide would fill it as fast as it was emptied. Twenty men would be set to pumping out the water while the remaining workers picked through the murk. With its miniature mountains and dips, the reef made digging difficult; it was hard to get a purchase on the rock and required hours of patience to make even the smallest indentation. As with the stones, Robert decreed that the pit should be shaped as accurately as possible to the eventual curves of the tower itself; one chip out of place, and the work was disrupted for days. He had reluctantly abandoned the possibility of blasting the foundations; dynamite was still too unreliable and could easily fracture the rock beyond repair.

Thus every chip and fragment had to be hammered by hand, taking days of work and skill. The masons also discovered that the deeper they dug, the harder and less workable the reef became. Robert, as usual, took precautions, and had a smith and forge already assembled at the rock so picks and boring-irons could be sharpened as hastily as possible. Conditions were easier while the weather was good, but in rain or strong wind, the workmen could only make progress at a snail's pace. Robert found himself striking an uncomfortable balance between speed and rigour; the day's work had to be done as quickly as possible before the next tide rose and covered the rock, but at the same time he could not risk the possibility of shoddy workmanship. However, by early July, the pit was considered ready enough to allow Robert to escort the foundation stone from Arbroath.

Much of the work revealed Robert's affection for ceremony. Everything, from the start of digging to the end of the highest tides, was accompanied by some form of fanfare. In part, it was his method of rewarding the men for their work. More than that, it showed his own militaristic leanings. Robert had a fetish for organisation; he liked his emotions well-disciplined and his joy neatly timetabled. Military discipline – brasswork, uniform, duty, honour – was also the most efficient method of ensuring a loyal and obedient workforce. Since the lighthouse service was still young, Robert could have tried almost any method of managing his employees; as it was, he chose the method which seemed most easily comprehensible to the men and which most suited him.

The naval flavour which he gave the service remained long after he was gone. He commissioned a special prayer 'for the use of those employed at the erection of the Bell Rock Light-House', petitioning God to 'prosper, we beseech thee, the work itself in which we are engaged. May it remain long after our eyes have ceased to behold it.' Hoisting flags, chanting blessings and nipping away at ceremonial drams were Robert's methods of expressing happiness he would not have stopped to analyse. In later years he regarded his time at the Bell Rock not just with pride but with nostalgia. In hindsight it became the point in his life when he had been most completely fulfilled. As with the arduous inspection voyages, the risks were part of the plea-sure. He kept the respect of his men as much through his own involvement in their work as he did through good management and consideration.

Nor was he indifferent to the peculiarities of his work. Even to Robert's prosaic gaze, the project often presented a strange, almost supernatural appearance. When the digging was at its height, the reef reverberated with the clamour of metal, fire and stone, casting out its elemental image over the seas. He noted on one particularly productive June day that

> The surface of the rock was crowded with men, the two forges flaming, the one above the other, upon the beacon, while the anvils thundered with the rebounding noise of their wooden supports, and formed a curious contrast with the occasional clamour of the surges ... In the course of the forenoon, the beacon exhibited a still more extraordinary appearance than the rock had done in the morning. The sea being smooth, it seemed to be afloat upon the water, with a number of men supporting themselves in all the variety of attitude and position: while, from the upper part of this wooden house, the volumes of smoke which ascended from the forges gave the whole a very curious and fanciful appearance.

Robert was not a man easily moved, but he found the image of man placing his improving mark on nature as satisfying as many of his contemporaries found it satanic.

Once the pit had been levelled, work began on the first few courses of foundations. Each stone had to be jigsawed carefully into place, checked for flaws, and smoothed off. Robert used the minimum of mortar and trenails (the small oak poles joining one course to another), having designed the dovetailed stones so perfectly that no single one could become dislodged without moving the others. Even with the pedantry that this entailed, the work progressed much more quickly after the foundations had been laid; once the first two courses were set, the pit was filled and there was no longer the need for constantly pumping water out. By late September, three full courses had been laid, but the weather was getting so bad that work had to be stopped for the winter. Robert left the rock well satisfied, though he made several excursions during the winter to check that nothing had been broken or dislodged. As before, Rennie made a brief seasick visit to the reef in mid-December and submitted his

report to the Commissioners. His role had now been so completely diminished that he could not do much except reiterate
what Robert had already told them and insert a few quibbling
remarks about the correct form of mortaring.

When work began again in May 1809, Robert concentrated
on completing the top floors of the beacon to allow for extra
accommodation for some of the workmen. With its six iron-
bound legs and its sturdy wooden tower, it looked much like a
child's model of a moon-rocket with a small blazing globe of
light on its roof. An iron walkway was built as well, connecting
it to the foundations of the lighthouse. By early June, everything
except the furnishings had been moved into place and eleven
men moved gratefully from the cramped *Smeaton* to the beacon.
A few nights later, the weather lurched towards hurricane force.
Robert and the remaining workmen stuck on the *Smeaton* clung
to the gunwales and waited for the storm to exhaust itself. Thirty
hours later, Robert was able to land on the rock and find out
how the stranded men in the beacon had fared. He was alarmed
to find that three of the stones from the foundation – each
weighing up to a ton – had been lifted from their places and
shifted sideways, and that the lowest floor of the beacon had
been washed out clean. The men, meanwhile, had clung to the
existing stores to prevent them slipping from the windows and
wrapped themselves in old sails to keep the water out. James
Glen, one of the joiners, had spent the night telling the group
about his exploits on a North Sea ship in which he had been
reduced to catching and eating the ship's rats to fend off star-
vation. The stories had a miraculous effect on the men, who
later complained only mildly of their ordeal.

Once building work began again on the tower, it rose quickly
from the foundation pit. The first twenty-five courses, Robert
had decided, were to provide a solid stub on which the lighthouse and its apartments would sit. That crucial thirty-one feet
of solid, dovetailed granite would give the tower both its flexibility and its strength; the sea could claw as much as it liked

at the foundations, but would encounter the minimum resistance from the curved stone. As before, the design was Smeaton's invention brought to Robert's perfection. By the end of June, they were at work on the ninth courses. Each layer took two days or more to lay and mortar into place, but as the stubby tower tapered upwards from its foundations and the joiners became more practised at their job, work speeded up. And, with the beacon now finished, the workers were able to stay on the rock for most of the day without needing to go back and forth from the Smeaton for food or rest.

Elsewhere plans did occasionally go awry. One morning, an urgent letter arrived from the foreman at the Arbroath workyard. All British ships were lying embargoed in port, he wrote; the state of politics and the imminent prospect of war with France meant that no ship could leave shore until permitted to do so by the government. Accordingly the local port officer had prevented any of the lighthouse supply boats leaving. With the building work at a crucial stage, Robert worked himself into a frenzy of impatience. When a plea from the Sheriff of Forfar to the Customs Board in Edinburgh did nothing to change the situation, Robert became even more testy. The Board declared that it was a matter for the Lords of the Treasury in London to decide and until they had passed judgement, Robert and the Bell Rock would just have to wait.

Robert seethed for ten days, and then tried another tactic. As he pointed out, the men needed provisions; surely a boat bringing food and tools would be permitted just a few stones? The Customs Officer considered and relented, Robert was allowed his few stones, and building work went on, albeit surreptitiously. Until the embargo was lifted, however, Robert was forced to rely on a much diminished workforce and the goodwill of the Port Officer, who, Robert noted slyly, helped 'mainly through the . . . liberal interpretation of his orders'. By the end of August when work stopped for the year, the solid foundations had been finished, the embargo lifted, and Robert was pleased. In one season, the light had taken 1,300 tons of granite and risen over thirty feet from the ground. That year, Rennie made no visit and did not submit his customary report to the Commissioners.

When Robert and the workmen returned to Arbroath, with the usual three-gun salute and hoisting of flags, he was clearly beginning to feel that the Bell Rock represented a way of life for everyone involved. He wrote delightedly that

> Nothing can equal the happy manner in which these excellent workmen spent their time. While at the rock, between the tides, they amused themselves in reading, fishing, music, playing cards, draughts, etc, or in sporting with one another. In the work-yard at Arbroath the young men were, almost without exception, employed in the evening at school, in writing and arithmetic, and not a few were learning architectural drawing, for which they had every convenience and facility . . . It therefore affords the most pleasing reflections to look back upon the pursuits of about sixty individuals who for years conducted themselves, on all occasions, in a sober and rational manner.

His own conduct had long ago earned the trust of the men. Robert rooted himself as deeply in the work as his staff, and for all the stiffness of his standards, they felt him to be an approachable character. When men were injured, he organised treatment, when invalided, he settled pensions, and when exhausted, he rewarded them with nips of whisky and offhand compliments. Their pay was still paltry – only 20 shillings a week, whatever the conditions – but they were reasonably content. These, after all, were skilled and able men, who understood the value of work well done as instinctively as Robert did. During the evenings, they sat in cramped intimacy up in the beacon, the beds laid five-deep up the walls, playing cards, writing letters and chatting. In fine weather, some would fish for the small fry surrounding the reef, and in bad, there would be fiddle music and singing. Robert had a small cabin to himself, little more than four feet broad, in which he kept a cot-bed, a folding table, a few books, a barometer and two small stools. Occasionally he received visitors there, but most of the time he was writing letters, keeping up his journal or 'making practical experiments of the fewness of the positive wants of men.' The Bible, he declared, was his only truly essential item.

Yet 1810 was to be the most demanding year so far. Work started in mid-April, with Robert determined to hasten matters on as swiftly as possible. The men found the beacon intact, but a little damaged by the winter seas; it had been used as a convenient roosting place by the local sea birds, who had daubed it with an unwelcome whitewash of guano. Inside, the rooms smelt musty and dank, but were only a little affected by seawater. The stump of the half-made tower was undamaged by the sea, and the masons set about fixing the first few courses of the walls and spiral stairs immediately. Despite storms that delayed work by several days, by early June they had finished the first floor. It was a fidgety task, often more like sculpture than building work, since it was necessary now not just to dovetail the blocks, but to ensure that their outer edges and the tapering gradient never unbalanced the shape of the building. In addition, Robert's insistence on using pozzolana mortar, which was stronger but took longer to dry, meant that the men spent their days playing a delicate game of grandmother's footsteps with the sea: wait too long for the mortar to set, and the chance to work disappeared as the tide rose; wait too little, and the damp mortar would be unglued by the salt water.

All work remained dependent on the weather. Robert grew astute at watching for signs of an impending gale, noting the habits of the sea birds and the way in which shoals of fish clustered over the reef in good weather but vanished into the deeps as bad weather approached. During patches of enforced idleness, he would sit watching the waves curl and unfurl against the walls of the tower, and study the way gales behaved. He made loose mental calculations of the weight of each breaker, the way each fourth wave would be mightier than the rest in a storm, and the eerie way in which the air moved around the water. His interest was, in part, detached scientific curiosity, but it was also the habit of a man accustomed to judging tactics. Some of the gales were indeed spectacular; the men ran the constant risk of being flung off the walls of the tower by the waves, many of which broke easily over the top of the seventy-foot top course. Robert, characteristically, was torn between worry for the safety of the workmen and panic at the prospect of stones being washed off the light. 'The loss even of a single stone,' he wrote nervously, 'would have greatly retarded the work.'

After a while, the men grew discontented with their rations, and complained to Robert that they weren't getting a proper allowance of beer each evening. Robert was unsympathetic, deciding crossly that their complaint was an 'unexpected and most unnecessary demand'. His initial response was simply to reiterate the list of their existing rations, and to remark that it seemed quite enough to him. The boat-master and another recently arrived workman replied that no change was no answer, and that they, and many of the remaining workforce, would come out on strike if nothing were done. Robert appealed to their humanity, reminded them that they would be sabotaging 'a building so intimately connected with the best interests of navigation', and threatened to fling them out to the local press gangs if they didn't return to work immediately. At the end of his speech, he told them that 'it was now therefore required of any man who, in this disgraceful manner chose to leave the service, that he should instantly make his appearance on deck.' Robert's threats worked. Most of the men returned, grumbling, while Robert dismissed the two ringleaders along with an indignant letter to the Arbroath Foreman. 'Nothing,' he wrote, 'can be more unreasonable than the conduct of the seamen on this occasion, as the landing-masters crew not only had their own allowance on board the Tender, but, in the course of this day, they had drawn no fewer than twenty-four quart pots of beer from the stock of the *Patriot* while unloading her.' Nothing more of mutiny or strike was heard for the rest of the works.

By early July, the tower was almost finished. The last few stones were being laid and the final journeys from Arbroath completed. Robert made plans for the ceremonial 'finishing pint', while the Arbroath stonecutters gathered their sweethearts for a dance. Robert took his first stroll around the balcony, feeling 'no small degree of pleasure' in his lofty exercise yard. The final stone – the lintel of the lightroom door – was laid on 30 July, with Robert adding his own ceremonial blessings. 'May the Great Architect of the Universe,' he announced, 'under whose blessing this perilous work has prospered, preserve it as a guide to the mariner.' Each of the lighthouse boats displayed their flags, and Bassey the horse was 'ornamented with bows and streamers of various colours'. Robert made a brief but heartfelt speech, reminding the men of their duties to commerce and hard work, and, with an emotional flourish, assuring all of them that he remained forever grateful for their contributions. The health and prosperity of the Bell Rock was drunk several times over with feeling and three days later the working party left for Arbroath.

Before the lantern could be fitted and the light displayed, the weather flung one final challenge at the Bell Rock. A storm blew up and lasted for four days, pounding the reef and making the building tremble. The waves reached well over the top of the tower (now 110 feet high) and cascaded down through the roofless apartments, shifting tools and terrifying the remaining workmen. Several mortar tubs and the smith's anvil were flung from the barracks down onto the rock, while the remaining men in the barracks huddled in corners as the rain seethed in on them. Once the storm abated, Robert, anxious to avoid a repeat in future, dismantled the bridge between the barracks and the tower and hastily completed the fitting of the light room. Glass, stone and water were an uneasy mixture, and Robert spent much of the next two months in a state of near panic waiting for the new lantern glass to be fitted up and ready. In this, as in many other aspects of the works, Robert designed according to circumstance. He had to find a new kind of glass, built to withstand heavy seas and sizeable enough to be seen from a distance. In front of the rotating reflectors, Robert intended to place panes of red glass which, when turned, would give the Bell Rock an alternately red and white beam. Up until then, no red glass of the right size yet existed. Robert, as usual, made it exist.

The final task was to appoint light-keepers. Robert chose two familiar hands: John Reid, who had worked on the floating light, and according to Robert's approving verdict, was 'a person possessed of the strictest notions of duty and habits of regularity from long service on board of a man-of-war'; and Peter Fortune, appointed as assistant keeper, who 'had one of the most happy and contented dispositions imaginable'. Soon after, Robert added a third keeper, since, as he explained to the Commissioners,

Two keepers are being considered for each of the other lighthouses, but on such a station as the Bell Rock there cannot be fewer than four keepers – three always in the house and one ashore at liberty – the third keeper is not so necessary on the actual duty to be performed but in the case of sickness or death – as happened at the Eddystone when there were only two keepers there. One died and the other, for fear of being suspected for murder, kept the corpse of his companion for about four weeks. When the state of the weather permitted the attending boat to go off to the lighthouse, the poor man was found sitting upon the Rock. The body of the deceased was in such a state of putrefaction that it was long before the effect of it left the apartment of the lighthouse.

The abandoned keeper had apparently maintained the light, the records and the watches perfectly, but when he was eventually brought off the rock, he was found to have gone completely insane. Until the 1990s, when all the lighthouses were automated by the Northern Lighthouse Board, each lighthouse therefore had three keepers 'to prevent suspicion of murder'.

The first Bell Rock keepers were less bothered by homicide than bad weather. The two lasted the winter alone on the rock, more in waiting than in usefulness, since the light could not be fully lit until the following season. When Robert did return in January, he found both in good spirits. Mr Reid mentioned that a couple of passing gales had made the tower shake beneath them, and though they had every confidence in its strength, 'he nevertheless confessed that, in so forlorn a situation, they were not insensible to those motions which, he emphatically observed, "made a man look back upon his former life".' Finally, on the first of February, the Bell Rock lighthouse was lit for the first time, sending a clear beam over the sea 'like a star of the first magnitude'. As the notice posted in maritime journals put it,

> A lighthouse having been erected upon the Inch Cape or Bell Rock, situated at the entrance to the Firth of Forth and Tay in north lat. 56° 29′ and west long. 2° 22′, the Commissioners of the Northern Lighthouses hereby give notice that the light will be from oil, with reflectors, placed at a height of about one hundred and eight feet above the medium level of the sea. The light will be exhibited on the night of Friday, the first day of February 1811, and each night thereafter, from the going away of daylight in the evening to the return of daylight in the morning.

Robert returned, swelled with success, to Edinburgh and home. The Commissioners emerged, one by one, to visit the rock and praise Robert's achievement. In the Arbroath workyard, the workmen celebrated and dispersed, many of them moving on to other less glamorous jobs for the Northern Lights. Back in Edinburgh, Robert was applauded in the newspapers, courted by the gentry and famed within his own profession. Robert revelled in the acclaim. His pride in his own work manifested itself in fussiness over the light; he fidgeted endlessly with stores and rations and lenses, and demanded a constant series of reports and letters from those based on the rock. To capitalise on the keen public interest in the works, the Commissioners suggested that Robert should begin preparing an account of the works for publication. He dithered, sorting through plans and correcting his journals. One duty he did attend to was a geographical plan of the Bell Rock reef, complete with carefully dedicated place names. The scrubby hummocks of rock gained a sudden grandeur, with names like Port Stevenson, Hope's Wharf and Smith's Ledge. Robert made sure not only to commemorate each of the Lighthouse Commissioners, but also many of the artificers who had worked on the rock. There was also a small patch of ground entitled 'the Last Hope', after the fearful escape the men and Robert had from drowning during 1807. Perhaps deliberately, Port Rennie was stuck some way out to sea.

Rennie himself did not feature heavily in Robert's reminiscences. Though he collected his chief engineer's fee of £400 from the Commissioners, the episode clearly rankled. In a letter of 1814 written to a friend, his usual goodwill shows signs of strain. The plans for the Bell Rock lighthouse, he noted, had been drawn up by him, though,

> When the work was completed, Stevenson considered that he had acquired sufficient knowledge to start as a civil engineer, and in that line he has been most indefatigable in looking after employment, by writing and applying wherever he thought there was a chance of success ... he has assumed the merit of applying

coloured glass to lighthouses, of which Huddart was the actual inventor, and I have no doubt that he will also assume the whole merit of planning and erecting the Bell Rock Lighthouse, if he has not already done so. I am told that few weeks pass without a puff or two in his favour in the Edinburgh newspapers.

Rennie's sour predictions came true. When he heard of the book Robert was compiling he was piqued, particularly since Robert had not bothered to contact him. The two had kept up a desultory correspondence for a few years after the completion of the works, though all warmth had long since been demolished. 'I have no wish,' Rennie wrote to a friend, 'to prevent him writing a book. If he details the truth fairly and impartially, I am satisfied. I do not wish to arrogate to myself any more than is justly my due, and I do not want to degrade him. If he writes what is not true, he will only expose himself. I bethink me of what Job said, "Oh, that mine enemy would write a book!"'

Robert's *Account of the Bell Rock Lighthouse Including the Details of the Erection and Peculiar Structure of that Edifice* only mentioned Rennie in the occasional polite aside when it was finally published. According to Robert, Rennie had been more of a disinterested patron than a paid-up participant. Though both Rennie and Robert went on to grander things, Rennie always felt slighted. His equanimity finally cracked some time later, when he was tentatively approached by the subscribers to the Stockton to Darlington Railway. They would, they wrote, be delighted if Rennie and Robert would consider jointly surveying the site. Rennie snapped back, 'If the subscribers to this scheme have not sufficient confidence in me to be guided by my advice, I must decline all further concern with it.'

After Rennie's death, his son Sir John Rennie published his own account of the Bell Rock works; in it, he asserted that his father had 'designed and built' the lighthouse. Four years later, he published a further history, emphasising again that 'the design was prepared by the late Mr Rennie; that no modifications were introduced without his sanction and consent; and that from first to last he was responsible for the success of the undertaking'. Samuel Smiles, in his essay on Rennie in the *Lives of the Engineers*, took Sir John's version, though he did concede that 'his name has not usually been identified with the erection of the structure; the credit having been almost exclusively given to Mr Robert Stevenson, the resident engineer, arising, no doubt, from the circumstance of Mr Rennie being in a great measure ignored in the "Account of the Bell Rock Lighthouse", published by Mr Stevenson several years after the death of Mr Rennie.' The dispute between the two sides grumbled on for several decades; the Rennies protesting their part in the works, the Stevensons publishing their own claims and disclaimers, the Rennies hitting back angrily. The Commissioners, meanwhile, responded with their usual languor, only conceding that to Robert Stevenson 'is due the honour of conceiving and executing the great work of the Bell Rock lighthouse', after Robert's death.

Rennie, however, was the lone dissenting voice. The paper war between the two sides was largely the preoccupation of vested interests; as far as Scotland and the world was concerned, Stevenson had built the Bell Rock and Stevenson, as he had fully intended, took the credit. Scotland, indeed, considered Robert a genius. He had achieved the impossible, lit up the darkened ocean and brought glory both to engineering and his country. He was praised for his upstanding interest in Britain's commerce, his disinterested contribution to the war effort with France and his generous advancement of the scientific cause. He was lauded as a true patriot, a mechanical prophet and a man in whom all the most noble qualities of the Romantic spirit were combined. As Rennie had bitterly predicted, the papers published almost weekly bulletins on his latest works, and the King of the Netherlands was so impressed that he sent a large gold medal in recognition of Robert's services to all European seafarers. Mr George Bruce of Leith composed a poem to Robert:

> The undertaking, oh! How vast, how grand,
> Which shall for ages as a monument stand
> To Stevenson's, a never-dying name,
> Wafted after by the loud trump of fame!

By 1810, Scott was synonymous with Scotland. His championing of the Highlands, his fancy-dress images of wild Jacobite life and his elaborate tales of noble clan feuds had seized the popular imagination. He was the first, and the most inspired, of Scotland's marketing men and his own life had become an indivisible part of his campaign. He was also a keen traveller; when the Lighthouse Commissioners offered him the chance to accompany Robert on the annual tour of inspection, Scott responded enthusiastically. He was pleased, as he wrote to a friend, 'to accompany them upon a nautical tour of Scotland, visiting all that is curious on continent and isle', and to make the acquaintance of 'the celebrated engineer, Stevenson'. 'I delight in these professional men of talent,' he confided to his publisher, only a little patronisingly. 'They always give you some new lights by the peculiarity of their habits and studies, so different from the people who are rounded, and smoothed, and ground down for conversation, and who can say all that every other person says, and – nothing more.'

The trip was, for both author and engineer, a success. After visiting the Isle of May, where Scott suggested the old lighthouse should be 'ruined *à la picturesque*', they reached the Bell Rock. Scott found it 'well worthy attention ... no description can give the idea of this slight, solitary, round tower, trembling amid the billows, and fifteen miles [sic] from Arbroath, the nearest shore.' Evidently entranced by the fable of Sir Ralph the Rover, he suggested that a mural should be painted on the keepers' apartments, complete with clanging bell and grim piratical drowning. Before leaving the light, one of the Commissioners asked him to inscribe their visitors' album. Scott took up his pen, mused for a while and then, to Robert's intense gratification, dashed off the following poem:

> Far in the bosom of the deep,
> O'er these wild shelves my watch I keep:
> A ruddy gem of changeful light,
> Bound on the dusky brow of night.
> The seaman bids my lustre hail,
> And scorns to strike his timorous sail.

Once back on the boat, a gale blew up and they hove-to for the night. According to Scott, everyone on board was sick, 'even Mr Stevenson'.

The party sailed up the east coast, anchoring occasionally to allow Robert to dash off and make his inspections. Scott spent his time eating, drinking and pondering the curious habits of the islanders. He also tried and failed to shoot a golden eagle, and was much impressed by the legends of the wreckers in Shetland. The diary Scott kept of the voyage was later to be expanded into a novel, *The Pirate*, and he seemed happy to discuss his literary inspirations with Robert and the crew. Many years later, Robert wrote his own account of the trip, berating himself for having taken so desultory an interest in Scott at the time. 'Had I been more fully *alive* to the eminence and ultimate celebrity of this "great and good man",' he apologised, 'I should have taken notes at the time.' As Robert lay on what was to be his deathbed, he felt the warm glow of mutual approval come over him again. Scott, he wrote, 'was the most industrious occupier of time ... he wrote much upon deck – often when his seat on the camp stool was by no means steady. He sometimes introduced Rob Roy's exploits in conversation so fully that when I read the Book many parts of it were like a second reading to me.'

Scott's interest in lighthouses soon paled. He left the more troublesome work to Robert and became preoccupied with the genealogies of neighbouring lords. His only quickening of curiosity came in August, when they were sailing round the Hebrides. He recorded that

> Having crept upon deck about four in the morning, I find we are beating to windward off the Isle of Tyree, with the determination on the part of Mr Stevenson that his constituents should visit a reef of rocks called Skerry Vhor, where he thought it would be essential to have a lighthouse. Loud remonstrances on the part of the Commissioners, who one and all declare they will subscribe to his opinion, whatever it may be, rather than continue this infernal buffeting. Quiet perseverance on the part of Mr S, and great kicking, bouncing, and squabbling upon that of the Yacht, who seems to like the idea of Skerry Vhor as little as the Commissioners. At length, by dint of exertion, come in sight of this long ridge of rocks (chiefly under water) on which the tide breaks in a most tremendous style. There appear a few low broad rocks at one end of the reef, which is about a mile in length. These are never entirely under water, though the surf dashes over them. To go through all the forms, Hamilton, Duff, [two of the Commissioners] and I, resolve to land upon these bare rocks in company with Mr Stevenson. Pull through a very heavy swell with great difficulty, and approach a tremendous surf dashing over black pointed rocks. Our rowers, however, get the boat into a quiet creek between two rocks, where we contrive to land well wetted. I saw nothing remarkable in my way, excepting several seals, which we might have shot, but, in the doubtful circumstances of the landing, we did not care to bring guns. We took possession of the rock in name of the Commissioners, and generously bestowed our own great names on its crags and creeks. The rock was carefully measured by Mr S.

> It will be a most desolate position for a lighthouse –
> the Bell Rock and Eddystone a joke to it, for the nearest
> land is the wild island of Tyree, at fourteen miles'
> distance.

Scott's distaste for the place was one day to prove prophetic.
But for the moment, Robert, having proved his point to the
Commissioners, was content to sail on.

Robert's opportunity to show Scott the lighthouse work was
the seal of the Bell Rock's glory. Scott's interest was proof of
Robert's success. He had achieved something that many had
considered impossible, and he was reaping his reward. As an
engineer and as an enterprising man at the centre of Edinburgh
society, he had arrived. Though far too pragmatic to rest or
gloat, he was well aware of the significance of his victory. The
Northern Lighthouse Commissioners now treated him with
respect and he was consulted by lords and politicians. At home,
his family was flourishing, and his children were starting to
show the rewards of their education. And, after years of patient
waiting, he had made his masterpiece. For a while at least, he
was a satisfied man.

FIVE

Edinburgh

Robert adjusted to his new status as grand young man of engineering with predictable ease. He was a man of standing now; a recognised figure with a position to maintain and a lifestyle to pay for. His reputation was made and his contribution to Scottish history assured. If he had wanted to, he could have retired then at the age of forty-two and lived the rest of his life in the knowledge that the Bell Rock alone would stand as his lasting testament. But Robert had no intention of retiring. He had completed one phase of his life but remained as restless as ever. As soon as he returned from the Bell Rock, he began his education again, still chasing the elusive degree and still convinced that he needed more science and more refinements to become his own vision of a 'man o' pairts'.

Given his faith in the practical sciences, it was unfortunate that Robert's first task was to finish off his account of the Bell Rock as commissioned by the board of the Northern Lights. Robert had accepted the commission but found when he came to it that he could not begin writing. True, he had plenty of material, including the records and Minutes of the Northern Lights, as well as his own copious journals, but he balked at turning them into print. In part, his shyness was a symptom of his old Achilles heel, his lack of 'book-learning', and his wariness of intellectual men. For all his new status, he still found book-ishness uncomfortable; his life was one of doing not thinking, and the idea of proving himself all over again in print was almost distasteful. The more time passed the more squeamish he grew.

He tinkered with his journals, dictated a few pages to his daughter Jane, now acting as his secretary, and sent sections off to Sir Walter Scott for comment. He commissioned J. M. Turner to provide an illustration for the book. Turner, without having actually seen the Bell Rock, produced an epic, storm-swept illustration which delighted Robert. He commissioned further engravings, and fussed over the records, but still could not begin writing. At Baxter's Place, he fell into long silences and moments of unexplained melancholia. After a couple of silent years, the Commissioners asked him to explain the delay. Robert wrote back, promising them that the book would be ready soon. He then contacted his old friend Patrick Neill, now a bookseller and publisher, who he hoped might goad him into action. Neill suggested that Robert should apply a little pressure to himself. 'Advertise it in the press,' he wrote, 'sell 300 to a bookseller – put the drawings in the hands of the engravers that they may draw on your purse – promise 50 copies to the Board, etc, etc. In short, commit yourself in every possible way – spurred in this way by the necessity of the case, you will get on excellently.' Robert did as he suggested, and then found himself using Neill's suggestions as a further excuse to prevaricate.

Finally, the Commissioners, who had long ago advanced Robert the money to research the book, demanded a finished copy immediately. Robert, impatient at his own inefficiency, babbled out reams of dictation to Jane. In 1824, thirteen years after he had begun, the book was finished. A proof copy, kept by Robert, still shows his scribbled corrections, dense and fastidious as a Dickens manuscript. The finished result evidently pleased him, since he fired off a letter immediately to the London publishers John Murray, demanding that each copy should be sold at 'not less than four guineas'. His estimation of his own significance, it seemed, had remained undamaged by the delay. 'I wish to free my hands of trouble in this respect,' he wrote. 'I would at the same time be understood as being impressed with a sense of the labour which the work has cost me – and with a desire as far as possible of being reimbursed. Considering the interest which the Eddystone Lighthouse has excited – though perhaps a less difficult work than the Bell Rock – and from the rising importance of the general subject of lighthouses – in proportion as the Naval and Mercantile Marine advances – it is not thought that so small an edition as 250 copies will be long in the hands of the Booksellers.' At the end of the letter, he added a postscript, pointing out, somewhat pompously, his list of professional appointments. 'I may further mention that if thought necessary or proper it might be noticed that I am a Member of the Antiquarian and Mercantile Societies of Edinburgh and Geological of London &c. and Engineer to the Commissioners of Northern Lights and Royal Burghs of Scotland &c. I will affix the only copy I have got of the dedication. Sir William Rae, Lord Advocate, took a copy of it the other day with him to London to get it presented in due form to the King.' The King, unfortunately, refused the dedication. Nevertheless, as Robert had predicted, the book sold well, becoming in its turn a gospel for all future lighthouse engineers.

Aside from the book, Robert had plenty to occupy himself with. While continuing his work for the Commissioners, who now gave him an assured annual salary of £400, he had resumed his private business as head of the Stevenson firm. His appointment as Engineer to the Convention of Scottish Burghs in 1813 gave him a steady income and allowed him to experiment with far more than dovetails and ocean gales. The position meant that he was now responsible for the supervision of almost all public engineering work for Scotland's towns and cities, for roads, bridges, canals, harbours and quarries, for railways, tramways and grand municipal statements. As with the lighthouses, he taught himself by working. Railways, for instance, were still in the most youthful stage of development, and Robert was responsible both for promoting their use and developing their technology. He was also credited with having anticipated John Macadam's work on roads by several years, devising a system of iron or stone track-lines laid on smooth sand and surrounded with cobbles. Until his efforts, and the work of Thomas Telford, almost all British roads, including city streets, were little better than rutted cart tracks. Robert, of course, considered their state a disgrace to Scotland, and did what he could to improve them.

Grandest of all his many projects was his work on Edinburgh. By 1814, as part of the city's attempts to restyle itself, the Nor' Loch had been drained and the trim Georgian grid of the New Town was almost complete. All it lacked was the necessary webbing of roads, bridges and paths to connect the New Town to the Old. Robert was brought in by the Convention to advise on much of this work, including devising a method of connecting the long span of Princes Street to Leith Walk. It was a project requiring both technical skill and aesthetic sensitivity. At present the road ended squarely against Calton's rump. Robert's solution was to continue onwards round the lee of the hill, and to construct a further geometry of roads beyond. To do so he had to demolish several existing buildings, dynamite the hillside and plough a route through Calton Cemetery, which contained the graves of his own children. As the judge Henry Cockburn later explained in his *Memorials*, 'The way of reaching Calton Hill was to go by Leith Street to its base (as may still be done), and then up the steep, narrow, stinking, spiral street which still remains, and was then the only approach. Scarcely any sacrifice could be too great that removed the houses from the end of Princes Street, and made a level road to the hill, or in other words, produced Waterloo Bridge. The effect was like the drawing up of the curtains in a theatre.' Robert considered several different solutions, finally, and characteristically, settling on the one that he felt would be good for the value of local property. As he pointed out in his 1814 report, it would also be most convenient for city traffic and would show off Edinburgh's plentiful assets to finest advantage. 'As a great addition to the individual comfort and convenience of the inhabitants of Edinburgh,' he wrote, 'the bridge over Calton Street will open an elegant access to the lands of the Calton Hill, from which the surrounding country forms one of the most delightful prospects of distant mountain ranges, detached hills and extensive sea-coast, with numerous ships ever plying in all directions together with the finest city scenery that is anywhere to be met with.' He was proud of his adopted city and in time became almost fatherly towards it. 'There is, perhaps,' he once wrote, 'no other city which presents so many attractions to the man of taste and of science as Edinburgh.' Waterloo Bridge, Regent's Road and London Road still link the three divided aspects of the city and still flaunt their views of sea, sky and city.

His Edinburgh improvements are also an illustration of Robert's peculiarly ideological brand of engineering. His belief throughout life was that all his works should, in some form, act as songs in stone to the greater glory of God and Scotland. The lighthouses not only showed his belief that mankind had a duty to preserve his fellow man, they were also an illustration of his pride in his country. His public works were all expressions of his sense that industry, in all its different aspects, would be Scotland's saving grace. Robert was an enthusiastic Unionist and an even more enthusiastic Monarchist but he was also eager to prove that the Scots could take on England at all its games, and win. He also believed in encouraging the same sentiments in those who worked for him. He was, in his own way, as much a social engineer as a civil or a marine engineer. He wanted, he said, to encourage others to rise as he had from nondescript beginnings through hard work to prosperity. Those who were lazy, shoddy or diffident deserved the meagre life they got; those who applied themselves patiently to work should, he felt, take their place alongside the best of Scotland's men. He started apprenticeships, set up pensions and paid for training for those disciples who he believed would be able to make the best of their opportunities. His schemes were, by and large, laudable, but sometimes they gave the impression of becoming production lines for an endless succession of imitation Robert Stevensons.

His rise coincided with one of the most radical periods in Scottish history – the Age of Improvement. By the 1820s, Scotland had become one of the great empire-building nations. While England provided the ideology for colonialism, the Scots, in their thousands, provided the skills and manpower to make it work. The Enlightenment had laid the essential foundations, allowing men of modest background to rise to prominence. Improvement merely provided the practical application to the rational theories. The early Victorians believed, as no other generation has before or since, that life, work and society could all be shaped to fit a particular politics. To tame the savage Highlander, they brought not just education but physical change as well: new houses, industry, roads and villages. To improve the Lowlands they planted trees, built more roads, brought in railways and steam. Factories were set up and filled with bewildered peasantry while agriculture moved away from old practices into new mechanistic ways of working. The universities expanded and began new scientific subjects; engineering, helped along by the pioneers, suddenly became fashionable.

Much of the impetus behind Improvement was benign and, considering the rigid thinking of only a century before, remarkably meritocratic. But in some places it brought its unfortunate consequences; such as emigration and the destruction of much of Highland society. The movement of the cleared Highlanders to the big Lowland cities also brought an unwanted burden to those industries trying to soak up the sudden influx of people. Robert's attitude to the Clearances was ambiguous; although he supported many of the theories of improvement, he had seen the emigrant boats himself and they had upset him deeply. He was one of the few Lowland men of his generation who had travelled widely in the Highlands and understood it on a level that most of his Edinburgh contemporaries could not. Instead of seeing the Highlanders as misguided savages who needed to be transformed, at best, into replicas of Lowland men, Robert was shrewd enough to take most men on their own merits.

Unusually, he respected Highland Gaelic culture and found many of the old practices sensible, not backward. He had worked with Highlanders over the years, appointed them as keepers, hired them as builders and foremen, and felt a much greater affinity with them than with many of his southern countrymen.

Robert nevertheless agreed with much of Improvement's stern doctrine. He believed in modern farming, benevolent landlordism and, with reservations, in bringing the Highlands closer to the Lowlands. He was also evangelical about education and industry; in Robert's opinion, there was no such thing as too much learning or too much work. He was, and always remained, a benevolent conservative. As his son David later pointed out, Robert 'had no taint of bigotry or party feeling', and was too canny to follow one political ideology to the exclusion of all others. His attitude to Scotland's landed grandees, who still ruled de facto over much of the country, was ambivalent. He liked and respected many of the old noble families, and was enough of a snob to flaunt their acquaintance, but found it almost puzzling that they did not, like him, believe in a meritocratic society.

The side of Improvement that least appealed to him was its effect on his country's image. This was the age of great Scottish myths, of the king in fictionalised kilt, of Walter Scott and Highland Societies. The real leaps made through agriculture, industry, roads and riches might have been useful, but they didn't quite catch the imagination as well as George IV in pink tights. And so, once the Stuart threat had been chased out of Scotland, the middle classes came back, cautiously at first, but then in increasing numbers. The English, and indeed many Lowlanders, found the notion of well-managed wildness satisfying, particularly when it was within a few hours' carriage ride and could be viewed from the comfort of a civilised castle. As industrialisation smoked the rest of the country into sameness, it was always reassuring to know that up there beyond the mountains noble savages still roamed.

Robert found the new tartanised image of his country strange. He was proud of Scotland's fashionability but bemused by the dewy-eyed musings of his New Town neighbours. In one of his occasional letters to Scott, with whom he maintained an intermittent correspondence, he recalled the author's fascination for all things Jacobite. He had noted, he wrote, that 'in conversation, you never term Prince Charles "The Pretender"' and was therefore sending on a small bag of articles, 'which I shall not attempt to describe', given to him by the eighty-four-year-old daughter of a clergyman who had been with Bonnie Prince Charlie at Culloden. Robert had found Scott's habit of turning the Highlands into picturesque fiction odd. Like most of Edinburgh at the time, he admired Scott's Waverley novels, but had seen enough to know that Scott's heroic images were nonsensical. Admittedly, Robert did make his own contribution to the change in thinking. His roads and bridges and railways provided the means for Scotland's heritage industry to thrive.

Improvement of his country was only part of the story. Robert's most persistent search was for improvement of himself and those around him. The success of the Bell Rock and his appointment to the Scottish Burghs produced ample new works for the Stevenson engineering business. The queues of petitioners seeking Robert's advice increased so much that he began complaining that Baxter's Place was too crowded with callers to allow him any peace. Much of the work he was being given was mundane – reports on harbour works, consultations on bridges or supervision of drainage – but Robert had a horror of turning work down. Instead, he took on apprentices who worked on the tedious detail while Robert consulted on more prestigious works. James and Alexander Slight, David Logan and James Ritson were all old and trusted colleagues who had worked with him for upwards of twenty years. Ritson in particular became a close family friend of the Stevensons and was regarded by Robert's children as their ally and friend. All Robert's apprentices were expected to pay their dues, taking winter classes and training for decades. Those who did pass his exacting standards would be given unprecedented authority to supervise building works and take on their own responsibilities.

In addition to his senior assistants, Robert also took in younger apprentices to the workshops. Baxter Street was almost becoming a miniature university for engineers. Fathers or guardians, casting around for something suitable as an occupation for their sons, wrote nervously at first to Robert enquiring just what exactly this engineering thing was. By the 1820s, the few diffident enquirers had swelled to a flood. One letter, to Sir Edward Lees, who was seeking a position for his son, stipulated Robert's conditions:

> I have to notice that the young engineer should be educated as for a liberal profession. Notwithstanding that your young friend may have completed his grammatical and arithmetical *school* education he must now attend for at least three years to the higher branches at College and other classes. Especially for Mathematics, Chemistry, Natural Philosophy and Natural History, to architectural drawing, hand sketching &c. Besides attending to the above branches my sons and the young men of my office were sent to a millwright's shop or a Foundry for 12 or 18 months before entering the office.

As the petitioners – rich or poor, grand or desperate – increased, so did Robert's demands on them. A full apprenticeship, he wrote to Patrick Syme, one of his old colleagues on the Bell Rock who was now looking for a position for his son, was 100 guineas for a full year's work, 'payable in advance for three years ... the office hours are from 10 till 4 and from 7 until 9 but we modify these while the [university] classes are sitting to enable our young men to attend them. We shall do what we can,' he added reassuringly, 'to make a good engineer of him.'

Robert's attitude to nepotism was straightforward; he encouraged it. The sons of the Bell Rock workmen were now looking for jobs and were welcomed in the Stevenson workshops. Much of the success of Robert's apprenticeships was due to his gift for self-advertisement. But much of it was also due to the gathering prestige of engineering itself. The well-publicised successes of the early pioneers – Telford, Rennie, Smeaton – had allowed the next generation to flourish. By now, many of the engineers were not just aware of each other's existence, but drew on each other's expertise. Networks were established, correspondence exchanged, scientific writings mused over. Robert could now draw on other experts' knowledge to produce solutions to many of the most pressing technological problems. This small self-enclosed community of iron and steam brought its own cameraderie. The Stevensons, first Robert and then his sons, corresponded with almost all the major names in British Victorian engineering, including Telford, Stephenson, Faraday and Watt. Robert met Thomas Telford briefly in 1826 while working on the Annan bridge, and later despatched his eldest son, Alan, to serve a brief road-building apprenticeship with the Telford & Walker firm.

He was particularly effusive over his connection with George Stephenson, *Rocket* scientist. While developing the first of his locomotives, Stephenson wrote frequently to Robert on the subject of railways. Both were interested in the practicalities of rail: what was the ideal gauge, what was the maximum gradient, what were the benefits of steam versus horses. Robert was so overcome by their possibilities that he believed that many roads would eventually become redundant or be replaced with cast iron tracks. 'It will be found an immense improvement upon the Common Road and also the Tram or wooden railway,' he wrote in a letter to a friend in 1818. 'It is not only intimately connected with Inland navigation and augmented with it; but will be found as it becomes more perfect still to add to the resources of the Miner while every progressive step must advance the immediate interest of the Agriculturalist, the Merchant and the Mariner and in short of the community at large.' Stephenson welcomed his interest, and gratefully confided to Robert in 1821 that 'I know you have been at more trouble than any man I know of in searching for the utility of railways.' In 1822, Robert tried to buy shares in the Liverpool to Manchester railway, but found to his disappointment that they had all been sold. Instead, he contented himself with a tour of possible new sites for railways, and in sending Stephenson peremptory letters.

Similarly, Robert kept in touch with James Watt, then working on prototype steam-driven passenger ships in Glasgow. The correspondence was brief but richly typical. Robert, as usual, spent much of it soliciting for business. While drawing up plans for the Edinburgh–Glasgow railway, it had occurred to Robert that there might be a satisfactory bargain to be struck between Watt's enterprise, the railway and a new harbour Robert was building at Granton. 'It appears,' he wrote in an 1836 letter, 'to be of much consequence to your Steam Packet establishment between this and London that the Edinburgh and Glasgow *should come to Granton* and that it should be brought by the most easy line of draught . . . I have heard that you are about to build another ship – if you think well of it, it would tend to bring Granton Pier into much notice and be a high compliment to His Grace were the company to name her "The Granton".' Watt, sensibly, chose to ignore Robert's suggestion and picked both his own route and his own name. The setback did not in the least discourage Robert, who simply turned his attention elsewhere.

While Robert was busy pursuing his landbound enthusiasms, the business of the Northern Lights ticked steadily onwards. The NLB had now been given parliamentary authority to build lights where and when it considered them necessary without the need to gain London's approval for each one. Robert had already planned many of the most urgent sites, and between

1812 and 1833 was responsible for eighteen new lights all around the coast, including Cape Wrath, the Mull of Galloway and Dunnet Head. All of them were island or mainland lights, built to a well-established template and considered relatively straightforward compared to the Bell Rock. With most of them, Robert could delegate much of the practical building work to his assistants and rarely needed to stay on site. He was also now responsible for lighting the Isle of Man, since Parliament had chosen to place the island under NLB supervision. The routine of Robert's life grew increasingly crowded. In addition to the annual inspection tours, he made one or two extra visits every year around the coast, instructing builders and fussing over the keepers. In Edinburgh, he spent his time soliciting for business, involved in existing projects and in staying up to date with the regular bureaucracy of the Commissioners.

While Robert was hopping around the coast spreading instruction and light, the state of the sea was also changing. In 1815, the press gang was finally abolished. With the end of the Napoleonic wars and the Treaty of Vienna, naval manpower slumped from 145,000 souls to around 19,000; those who remained in the navy now did so voluntarily. With the reduction in numbers, payment of wages improved and there was no longer the same enthusiastic reliance on rum, sodomy and the lash. With the absence of war came other changes. Steam began slowly to replace sail. Some form of training was introduced for seamen and the slave trade was finally halted. In 1822, the same year George IV arrived to inspect his northern kingdom, the last pirate in Scotland was hanged in Leith. In England, piracy hobbled on for a few more years, though it was not until 1840 that the last pirate was staked to the ground at Wapping's Execution Dock to wait for the rising tide.

The first stirrings of concern for mariners' conditions also began to produce effects. In 1836, a parliamentary select committee was established to 'inquire into the cause of the increased number of shipwrecks with a view to ascertaining whether such improvements might not be made in the construction, equipment and navigation of merchant vessels as would greatly diminish the annual loss of life at sea'. At the time, the mortality rate from shipwreck was running at around 1,000 deaths a year. As the Committee discovered, most of those deaths were preventable. Poor construction and design, inadequate equipment, overloading, masters' incompetence, lack of harbours, inaccurate charts and drunkenness were all rife within merchant shipping at the time. In the next few years, campaigning by Samuel Plimsoll forced the introduction of compulsory load lines (marking the level to which a ship could safely be weighted), merchant masters and mates were compelled to carry certification and the first attempt to lay down 'rules of the road' had been made. On their own, none of the changes made much of an impact on the figures, but cumulatively they marked the beginnings of a practical response to the mistakes of the past.

Robert took a keen interest in the changes in legislation and safety. He was years ahead of his time in recommending that ships should be properly lit, with a lamp on both port and starboard sides to indicate the width and proximity of ships travelling at night. He also kept a weather eye out for developments in England and abroad. In amongst one of his scrapbooks was a *Times* notice of September 1805 advertising *A Curious Aquatic Exhibition: Daniel's Life Preserver*. The life preserver was a forerunner of the cork life jacket, a waterproofed leather vest inflated by blowing into a silver tube. The notice announced that

> Several persons, thus equipped, jumped into the Thames from Mr D's boat ... and floated through the centre arch of London Bridge, with perfect ease and safety; some were observed to be smoking their pipes, others playing the flute and french horn, and which they seem'd to perform with as much indifference as on dry land ... The advantages which individuals thrown by accident or by shipwreck into the water, must

derive from being thus much on the surface are very great ... we sincerely congratulate the inventor on his ingenious discovery.

Robert also wholeheartedly approved of Manby's mortar line and was instrumental in encouraging its use throughout Scotland, complaining several times in his reports that its use was not more widespread. Not all of his interest in lifesaving was entirely altruistic, however. He compiled a thorough list of British harbours of refuge, pointing out that many had become inadequate or downright dangerous for the new steamships now in use and hinting that the Stevenson firm would be more than pleased to submit plans and a quote for any new harbour the local authorities might have in mind.

Harbours, lights and locomotives were not Robert's only pre-occupations. As his children began to grow, his family was now taking up as much time as the lighthouses. In 1814, after living out a comfortable old age at Baxter's Place, Thomas Smith died. In his final years, his main preoccupation had been his loathing of the French. Thomas, like much of Edinburgh at the time, feared the threat of revolution with a passion. As he grew older, his views became more extreme. 'The people of that land were his abhorrence,' Louis wrote later, 'he loathed Buonaparte like Antichrist. Towards the end he fell into a kind of dotage; his family must entertain him with games of tin soldiers, which he took a childish pleasure to array and overset; but those who played with him must be upon their guard, for if his side, which was always that of the English against the French, should chance to be defeated, there would be trouble in Baxter's Place.' He felt his son and three daughters comfortably provided for and entrusted them to Robert's care, though by the time of his death, Robert had already taken over his position as head of the household in all but name.

By 1820 Robert's eldest son, Alan, a delicate, thoughtful boy, was completing his time at the High School, while surreptitiously maintaining a furtive interest in literature and the classics. His second son, Bob, now entering his teens, showed no curiosity about lighthouses and only the most desultory interest in learning at all. David and Thomas were still too young to have learned much independence, though Robert had spent much of his time impressing on all of them the attractions of an engineer's life. Jane, meanwhile, was now working as Robert's assistant and secretary, writing letters, taking dictation and screening the queues of petitioners while Robert was away from Baxter's Place.

Home life was arranged according to Robert's ideal. The five children occupied their time with his vigorous notion of upbringing: fresh air, Scripture and study. Louis, writing over eighty years later, saw it as a time of pleasurable dread. He wrote:

> No 1 Baxter's Place, my grandfather's house, must have been a paradise for boys. It was of great size, with an infinity of cellars below, and of garrets, apple lofts, etc., above, and it had a long garden, which ran down to the foot of the Calton Hill, with an orchard that yearly filled the apple-loft, and a building at the foot frequently besieged and defended by the boys, where a poor golden eagle, trophy of some of my grandfather's Hebridean voyages, pined and screamed itself to death . . . Within, there was the seemingly rather awful rule of the old gentleman, tempered, I fancy by the mild and devout mother with her 'Keep me's.' There was a coming and going of odd, out-of-the-way characters, skippers, light-keepers, masons, and foremen of all sorts, whom my grandfather in his patriarchal fashion, liked to have about the house, and who were a never-failing delight to the boys. Tutors shed a gloom for an hour or so in the evening, and these and that accursed schoolgoing were the black parts of their life.

Robert, with his fear of wasted life, tended to treat his children much as he treated his employees, inflicting discipline through industry, and morality through example. He was, as Louis noted, an austere patriarch who believed that the best gift he could give his children was to mould them as he had moulded himself. In part, he was only conforming to the prevailing notions of childhood and youth. Like many of his contemporaries, he believed that mollycoddling one's children when they were young only led to weakness later. But this reflected more than just the habits of the day; it was also Robert's conviction that his children should be like him, think like him, work like him and love like him. Robert's approval was enthusiastic but sparing, and his children remained aware that his favour was conditional on good behaviour. If he wished to compliment them, he gave them a harder, more responsible task to perform. Love, like everything else in Robert's life, was something that must be earned. Mostly, they looked to their mother for the softer side. Their letters to her are more open, more honest, and, frequently, filled with worries that they knew Robert would not have tolerated. Louis records an early incident when Thomas escaped a dull family outing and wandered down to Portobello. Robert was more than just angry at Tom's disappearance, he was shocked. 'Long after,' writes Louis, 'my grandfather, who was off upon his tour of inspection, wrote home to Baxter's Place in one of his emphatic, inimitable letters: "the memory of Tom's weakness haunts me like a ghost."' Tom's crime hardly merited such a profound reaction, but it shows up Robert's terror of waywardness. His censorship was kindly meant, but it isn't surprising that his children grew up terrified of him.

At the age of eight, each of the boys was packed off in grand New Town style to the Royal High School, then at its old location on Infirmary Street. Once there, they were given a sturdy diet of classics and natural philosophy. The High School had already earned its reputation as Edinburgh's smartest public school; Sir Walter Scott and Henry Cockburn were both sent there in the 1790s, and it was still a source of irritation to Robert that he had not. By the 1820s, the school was moving from a diet of pure classics to a more practical curriculum. There was increasing emphasis on mathematics, science and physics, and the school now aimed to produce men of technical skill as well as intellectual learning. The determination to give the school a vocational role disturbed the old guard so much that by 1824 Henry Cockburn and Walter Scott established the Edinburgh Academy with the intention of restoring the balance back towards the classics.

For all its alleged usefulness, none of the Stevenson boys much enjoyed their time at the High School. Even Alan, by far the brightest of the children, did not shine. Bob, Robert's second son, left as soon as he could, having maintained only a hapless inability to spell. 'For this Journall, we are chiefly indibted to the persiverance of my Father,' he wrote in his diary the year he left, 'who on all convenuant ocastions aloted part of the day to sugest the proper subjects of observetion in hopes that in our future journeys or purshuts of life wi might keep up and observe the same practice.' Horrified, Robert sent Bob off for several crash courses in literacy, and then to St Andrew's University. Tom, the youngest child, was equally lacklustre. As Louis put it, 'Robert took education and success at school for a thing of infinite import; to Thomas, in his young independence, it all seemed Vanity of Vanities . . . Indeed, there seems to have been nothing more rooted in him than his contempt for all the ends, processes and ministers of education. Tutor was ever a by-word with him; "positively tutorial," he would say of people or manners he despised.'

From the moment Robert returned from the Bell Rock in 1814, his hopes rested mainly on Alan. His eldest son was a silent, introspective child, with an abstracted quality to him that Robert found disconcerting. His childhood had been marked by long intervals confined to bed. He, like his sickly nephew

Louis later on, spent his time reading and dreaming. By the time he reached the end of his High School years, it was evident to many who knew him that Alan simply didn't have the physical stamina that engineering required. He was calmer than his three rough-and-tumble younger brothers, and would often leave their games to retire quietly with a volume of poetry. By the time he was sixteen, he had developed into a character who stood a little to one side, with a self-sufficiency that few could penetrate. He participated in all family matters, helped his mother and plotted with his brothers, but he had an aloofness and an intellectual distance to him that the others could not reach.

Robert was encouraged by Alan's intelligence but disturbed by his solitary habits. When Alan showed an enthusiasm to continue studying the classics, his father grew fretful. Robert had never made any secret of the fact that he wished his sons to follow him into his profession. All the expense and trouble over educating his children was aimed almost solely at producing the next generation of Lighthouse Stevensons. Robert found Alan's frailty bewildering, since he himself had never had a day off work in his life and was not inclined to tolerate weediness or special pleading in others. So Robert cajoled and bullied Alan into shape. He wrote letters and dragged his son in for fraught, painful audiences, insisting again and again that Cicero or Pliny might be fine for men of leisure, but that status and security lay in practical skill. What use was a man who could quote Catullus in the original if he couldn't also build a well-laid road? As far as Robert was concerned, imparting any knowledge that couldn't be applied during a force-9 gale in the loneliness of the Atlantic was as much use as teaching dressage to dogs. At the heart of Robert's narrow attitude was the old black terror that, for all his accomplishments and all his fame, he was still a callow man, whose reputation depended on physical effort rather than intellectual brilliance.

*　　*　　*

And so Alan grew up aware that the weight of Robert's ambition rested on him. For years, he remained torn between following his talent for literature and his inevitable destiny as an engineer. In 1823, when Alan was sixteen, Robert wrote demanding that he should make a final decision. Two years into his time at Edinburgh University and after much thought, Alan finally replied, half teasing, half desperate for a resolution to his father's pressurings. He wrote in pinched schoolboy style,

> Dear Father,
> I take this opportunity of answering your letter dated – day of February, in which you stated a desire that I would apply myself to some *business*; and although, I must confess, I had a liking for the profession of a soldier, on the receipt of your letter I determined to overcome this foolish wish and am happy to say I have succeeded. On further consideration I found in myself a strong desire of literary glory, and I pitched upon an advocate but there was want of interest. I was the same way with a clergyman; and as I am by no means fond of shopkeeping I determined upon an engineer, especially that all with whom I had spoken on the subject recommend it, and as you yourself seem to point it out as the most fit situation in life I could choose. I only doubt that my talents do not lie that way, but in hopes that my choice will meet with your full approbation.

The letter is partly a self-mocking reference to his own swither-ings, but it is also the forlorn plea of a boy who merely wanted to please his father.

Despite Alan's reservations about his suitability for engineering, his decision delighted Robert. Alan was packed off almost immediately after completing his university studies to London. He was to stay with the Reverend Pettingal in Twickenham, to be 'finished' according to the refinements of the day: learning dancing, touring England and polishing up his knowledge of gentlemanly habits. Naturally, his time in the south had a practical application as well. Alan was sent to several engineering works and spent his evenings studying yet more mathematics. Four months after his decision, he seemed to be submitting to his duty with grace and dry humour. He wrote from Ryde,

> I have little to say of ill habits, for Mr Pettingal has advised to wear gloves in the house to prevent me from biting my nails and scratching my fingers . . . I am going on with Leshe's *Analysis* and you may tell Mr Noble that I write out and repeat two Propositions every day which takes me little less than two hours. I think I shall be able to get through the end of the *Analysis* by the end of September or so; (Books studied include Wood's *Algebra*, Wallace's *Elements of Plain Trigonometry and Doctrine of Proportion*, and Brown's *Logarithms*) . . . I am getting on with Hume's *History* and finishing *Henry VII*. You may tell Mr McGregor that I shall attend to the difference between 'shall' and 'will'.

Alan's main comfort during his crash course in deportment and syntax was his cousin by marriage, Thomas Smith, the grandson of Robert's old lighthouse mentor. He had also been sent by Robert to Twickenham to be finished, and evidently found a strong dose of absurdity in their studies. 'Alan gives me books and newspapers to read,' he wrote in a letter to Robert in December 1823, 'and sometimes when Mrs P and Charles have gone to their rooms sends for me from the study to come and play at draughts or backgammon.' Thomas went on to Pembroke College in Oxford the following year, and thus remained for most of his time in England. Alan, his finishing finished, returned obediently to Scotland and to the next phase of apprenticeship to Robert.

Robert's plans for educating all his sons followed the same formula as his own apprenticeship with Thomas. Summers would be spent learning practical engineering at various sites around the country, winters were spent picking over theory both at home and at further university classes. Much of Alan's time was passed in Robert's office, copying reports, transcribing letters, writing out invoices or checking specifications. As Robert's workload had increased, so too had the accompanying bureaucracy. The older trainees and Alan's sister, Jane, helped with some of the snowfalls of print, but much of it still had to be dealt with by Robert. Once Alan arrived, however, Robert handed on much of the mundane work to him. Given that Robert's letter-writing habits had not diminished in the slightest – he still regularly filled 600-page letter books in less than six months – the sheer weight of material to be processed and recorded would have been enough to employ Alan for the rest of his life. But Robert wanted his son to become a complete engineer, not just someone capable of attending to office bureaucracy.

Alan also spent a great deal of time outside Edinburgh, inspecting the lights and learning the various disciplines of civil and marine engineering. Bridges, harbours, drainage, break-waters, roads and railways each had to be individually studied. During 1824 he was despatched to bridgeworks at Annan, har-bourworks at Crail and to the lighthouse under construction out on the Rinns of Islay. For a while he studied under Thomas Telford, examining riverworks down on the Wirral. Alan's introduction to Telford had, as usual, been organised by Robert, who felt it would benefit his son's future career to have studied under such an illustrious name.

Alan's submission to his father's chosen profession was almost complete; having made his decision, albeit under pressure, he learned swiftly. Only occasionally, after an unusually brusque instruction from Robert, did he show signs of chafing against his fate. At one stage, while in Hull supervising bridge works, he hinted that he might be able to find further work there, and seemed reluctant to return to Edinburgh. 'I should wish you to observe,' wrote Alan nervously to his father a little later while in Annan, 'that Edinburgh university classes start on the 25th October so that if I am to attend I should require to leave this on the 24th at latest.' Robert, who had already planned Alan's life for the next six years, seemed entirely deaf to the possibility that any of his children might need to live a life beyond lighthouses.

While Alan began his long climb to professional success, his brothers were catching up. Bob, by now crammed full of all the correct spelling he could ever need, was sent first to further classes in architectural drawing and French, before being posted down to London to improve his manners with Mr Pettingal. He, like Alan, had received a formal letter from Robert demanding to know what he should choose as a profession, but Bob wavered, aware that engineering was not what he wanted to do. By the time he returned from Twickenham, his dancing and his handwriting tweaked into line, he had made up his mind. He announced to Robert that he wanted to become a physician. He had been helped in this by Jane's newest suitor, a neighbouring doctor named Adam Warden who had encouraged Bob's interest. Robert, though disappointed, could not complain too hard since he already had one son marked down for engineering, and Bob was at least choosing a practical profession. He could do no more than give Bob his blessing and enrol him in further classes at university.

David was more encouraging to Robert. He, unlike Alan and Bob, showed no faltering in his enthusiasm for the Stevenson trade. He did moderately well at school without showing any sign of having picked up Bob's waywardness or Alan's poetic intelligence. He was also physically tough, and by his mid teens had become a sturdy boy without much originality of thought but with all the self-discipline needed for engineering. By 1830, having reached the age of sixteen and the end of his school career, David was taken along with Tom (now twelve) for the first time on Robert's annual voyage of inspection. The tour included all the Scottish lights as well as a brief detour to Wales and the Isle of Man. During the journey, David kept a journal, which is remarkable both for its length, 13,000 words, and its dullness. He had picked up many of his father's mannerisms, including Robert's fascination for architectural detail and his penchant for collecting statistics, but had not yet learned Robert's capacity to give them life and meaning. Most of this and his subsequent diary is written in the strained manner of a young gentleman on his first grand tour, desperate to brandish his cleverness and terrified to show enthusiasm. For the lack of anything else to say, he kept a faithful record of the prayers said on ship, the sermons from the local ministers and the local population statistics.

David also wrote a meticulous account of the Bell Rock, describing precisely the different rooms, the building materials and the method of storing water. 'The sixth floor forms the lightroom,' he noted, 'where the reflecting apparatus is ranged upon a perpendicular axis and made to revolve by a train of machinery driven by a heavy leaden weight. The reflectors are ranged in 4 faces 5 upon each face. They have beautifully polished faces of silver formed to the parabolic curve, each being illuminated by an argand burner.' While up in Bara, he brought the same care to his record of the islanders. 'They are very ignorant and cannot speak a word of English,' he wrote disdainfully. 'None of them can read a word in their own language, and – wonderful to say in this islet of the British Dominions – there is not among the natives a single leaf of a printed book, they are all Roman Catholics, and the priest pays them a visit once a year. Most of their houses consist of holes dug in the earth ... We went into one of them where we found several women and a number of children *squatting* upon the ground round a fire of peats burning on the floor.' David, like most of his Lowland compatriots, found Highlanders distasteful. Robert's respect for the Gaels had yet to filter through to him.

Once back in Edinburgh, David responded instantly to

Robert's formal letter. Yes, he agreed, he did want to become an engineer. Indeed, he had never considered any other profession, and would start his training as soon as Robert saw fit. He too was therefore despatched to Edinburgh University to take the usual diet of chemistry, mathematics, natural philosophy, architectural drawing and natural history. Robert, burned by Alan's example, also took the precaution of insisting that David took no classes in Latin or Greek. David succumbed happily. By 1832, he had already begun work on some of Robert's more prestigious projects, including the new bridge in Stirling and Robert's grand span of the Clyde, the Hutcheson Bridge.

As with Robert's work on Regent Road in Edinburgh, the Hutcheson Bridge was a prestigious commission, designed both as a replacement for a rickety predecessor and as a sign of Glasgow's industrial wealth. David was employed to help the masons and the foremen and was expected to carry out heavy manual work as well as drawing and planning. Robert's long training ensured that all his apprentices, particularly his sons, would learn the harder side of engineering. Not only were they expected to endure the long nights in cold lodgings or tedious days wetted with salt spray, they were also expected to be able to lift stones, carry water or lay foundations. Above all, as future managers of men, they were expected to look to the comfort and wellbeing of others before they looked after their own. Robert's policy paid useful dividends; by the end of the works on Stirling Bridge, the workmen had watched David change from his distant position as the chief engineer's son into a colleague who had rolled up his sleeves and laboured as hard as they had. Having earned their respect, they paid him back in trust. When Robert arrived to key the last stone of the bridge – with all the customary flag-waving, chain-wearing, speech-making ceremony of Robert's grandest works – David found himself elevated to the place of honorary team-mate. 'At six the workmen sat down to dinner in the open air,' he wrote excitedly, 'after which there was dancing to the bag-pipe. Mr Ritson and I were seized by a deputation from the men and carried *shoulder high* from the town across the service bridge to the green on which they were dancing where we were loudly cheered by the men and three or four hundred spectators.'

Part of his summer duties included work on the new road to Kintyre lighthouse. Out there on the gristly spit of land overlooking the Ulster coast, David also learned the isolation of lighthouse work. The light at Kintyre was one of the first to be constructed by the NLB back in the 1780s on a remote patch of moorland fifteen miles from Campbeltown. The light was halfway down a steep cliff which was inaccessible to boats and all materials, and supplies for the light had to be brought on horseback along a route snared with spating burns and hidden potholes. The road had been necessary for some time but, like much of the less urgent lighthouse work, it had been shelved until money permitted. David was charged with supervising much of the construction work, the laying of durable foundations and the purchase of stone and materials.

David wrote occasionally to his mother, confessing he was lonely. His lodgings were damp and uncomfortable, the workmen grumpy and he longed to be home. David did not admit his fears to his father. Robert, it was clear, was not interested in self-doubt and would have been furious if any of his children had confessed to feeling inadequate. 'I am coming on here very well,' wrote David cheerily to his father in August 1832. 'I have been levelling and surveying and also attending the road works and wishing myself sometimes with the *spade*, sometimes *breaking stones* and sometimes *quarrying*!!' The only hint that David was not entirely happy came in a final aside. 'I was glad to hear of Bob's safe arrival [in Edinburgh] – he will be with you before this reaches you. Give him my compliments and tell him I have been dreaming about him every night for the last week or 10 days. I am wearying to see him ... I hope to hear from you soon – let me know how long I am likely to stay here – I have been here four weeks already.' To make matters worse, he found himself in the middle of an outbreak of cholera. One of the workmen suddenly sickened and died and, when David next went to Campbeltown, he found the streets silent and the houses deserted. The locals, panicked by the prospect of a full-blown epidemic, had fled.

If Robert thought he could relax his vigilance after David and Alan joined the family business, he was mistaken. Thomas, it seemed, was proving to be as intransigent as Alan and Bob had been in their time. As the youngest child of the family, he had spent his youth being alternately indulged and ignored by parents exhausted by years of child-rearing. He was boisterous, full of tricks and eager spirits, victimised and adored by his older brothers. He, unlike Alan, had not had to bear the full weight of Robert's expectations, and had therefore been allowed much greater freedom to develop as he pleased. True, he too had gone through the same long pupilage of lighthouse tours and paternal sermons, but it was his brothers who had taken the brunt of Robert's zeal for education, discipline and industry. As a result, he lacked both Alan's intensity and David's pedantry. He was, and remained, a more spontaneous character than his brothers. His letters to Robert could be chattily informal in a way that Alan's had never been, and his habits had an exuberance to them that even Bob had lacked. As he grew older, he became a more complex character. On the one hand, he was more out-wardly demonstrative than his elder brothers, on the other, he was prone to bouts of debilitating melancholia and self-doubt.

As his son Louis later gleefully pointed out, Thomas's school years were not his finest moment. In fact, his experience seemed to consist mainly of being strapped for various crimes and inade-quacies. Tom was also at the High School when it finally moved from Infirmary Street to its grand new position on Regent Road, though his attitude to this sudden raise in status was soured by his lifelong loathing for school and schooling. Instead, he developed a desultory interest in books, natural history and literature. Louis later recalled,

> He had a collection of curiosities. He had a printing press and printed some sort of dismal paper on the *Spectator* plan, which did not, I think, ever get over the first page. He had a chest of chemicals, and made all manner of experiments, more or less abortive, as boys' experiments will be. But there was always a remarkable inconsequence, an unconscious spice of the true Satanic rebel nature in the boy. Whatever he played with was the reverse of what he was formally supposed to be engaged in learning. As soon as he went, for instance, to a class of chemistry, there were no more experiments made by him. The thing then ceased to be a pleasure and became an irking drudgery.

Once out of school, Tom seemed unenthusiastic about settling into any profession. Robert, as usual, wrote formally to him, demanding that he decide his business. Tom vacillated and then replied listlessly, saying that he had no particular career in mind just yet. Robert, exasperated, hauled him into the Baxter Street office to help with dogsbody duties. Once there, it was evident that Tom showed no more excitement about engineering than he had for anything else. Robert took him off on surveying expeditions and got him to work on planning and drawing, at which Tom proved more a hindrance than a help. At one point, he announced an interest in becoming a publisher. Robert helped him set up a working printing press and watched as Tom grew frustrated and then abandoned it. Then he said he would like to become a bookseller, so Robert took him to London to pick up tips from the local traders. That, too, came to nothing. Robert sent Tom back to Edinburgh to work for his old friend Patrick Neill, who had published Robert's Bell Rock book a few years back. He also wrote to Tom, hoping still to deal with him by force of persuasion. What Robert said in that letter was as revealing about his own philosophy as it was about Tom's behaviour. He wrote in August 1835,

> You were anxious to know the nature of the Bill Charges, and if it will have any tendency to stamp upon your mind the value of money – and its indispensable possession in this world, I shall think the double postage well bestowed . . . If you want to live as a gentleman, you must work as a man, for there is no dining without a purse. Now as you have not been born to the purse, you must just look after yourself. I blame myself for not sending you straight from the High School at once to business in some way. I regret my judgement in it with a young man who is perhaps not aware of the ground on which he stands. A young man who is perhaps more taken up with appearances than realities – who wants to be a gentleman without the means – who perhaps thinks of an exterior as the only thing . . . Consider the absolute necessity of making provision for the time when it will be asked of you, what is this man? Is he doing any good in the world? Has he the means of being 'one of us'? I beseech you Tom, do not trifle with this until it actually comes upon you. Bethink yourself and bestir yourself as a man.

Robert's letter only had the effect of making Tom stay on a little longer with Neill. When a chimney pot in the printing works crashed to the ground just beside him, Tom took it as a timely warning that publishing was a deadly profession, and left. Finally, having exhausted all possible idle options, he reappeared on his father's doorstep and asked him for an apprenticeship. In fact, Robert had little need for an extra apprentice at the time, since both Alan and David were now working full-time and he also had an existing stable of well-trained assistants. Nevertheless, it seemed like a useful settlement to his youngest son's dilettante habits and Robert accepted him. By spring of

1836, Tom had started classes at Edinburgh University and begun the long slog to full qualification, though he seemed as lacklustre over his new career as he had about his old ones. Meanwhile, sifting through the pockets of an old coat, Robert discovered something that seemed to justify all his darkest fears about Tom. Tucked into a corner out of sight was a small bundle of dog-eared papers. Tom, it seemed, had been secretly preoccupied with fiction.

If Tom had been caught smuggling gunpowder or dealing in wreck, the effect could not have been worse. Robert was so shocked by the discovery that he could scarcely bring himself to mention it. Finally, he wrote to Tom, painfully detailing the 'seven pages of my good card paper filled in your handwriting with great nonsense . . . I made an attempt to read it but I could not go on with it. I spoke to David about it. He looked at it and said there was a drawer full of such stuff in your room. I beseech you, Tom, give up such nonsense and mind your business. You are most ignorant of the history of your country and the science of your profession. As I have told you before – to little purpose I fear – this is not the time for you to write but to read lessons in morals and the practical details of your business. I have reminded you that I am 66 years of age and what then? I leave you to consider this and fill up the blank for yourself.' All Robert's old terrors returned. All his fears of producing literate but useless children were apparently justified. All the lessons learned by Alan's example, it seemed, had come to nothing. He need not have worried. Tom, mortified by his father's reaction, caved in almost immediately. Since it was plain that he would never make an author and that Robert, driven to extremity, was now applying the emotional thumbscrews, he flung away his attempts at writing, and finally applied himself properly to engineering.

For all Robert's warnings to his sons that he was getting older, he showed little evidence of reducing his usual hectic pace. If he felt the onset of old age, his actions scarcely betrayed it. Where possible, he still embarked on the annual circuits of Scotland for the lighthouse inspection tour, and spent much of the rest of the year darting from place to place, consulting and opining. He was often called down to London to appear in Select Committee proceedings as an expert witness and to appear at the meetings of the Society of Civil Engineers, of which he was now a member. Thanks to the Bell Rock, he was now moderately well known in the south and found himself well respected there. He, in turn, had an ambiguous affection for London. He liked its opulent streets and its sense of prosperity, but found English scenery dull and London habits stultifying. England, he emphasised over and over again, had little except government to offer Scotland. During the annual inspection tour of 1814, he wrote from Shetland to a friend, delightedly comparing north with south. 'I can assure you I felt no small satisfaction in comparing the aspect of the Northernmost with the most Southern point of the land of this happy Kingdom. For about this time last year I was at the Land's End, and I am fully of opinion, that the preference is due to Shetland whether considered for pasture or for Crops when compared with Cornwall.'

In part, his unrelenting pace was a symptom of Robert's old fears. He remained terrified of being unable to provide for himself and his family. Like many self-made men, he remained haunted by the remembrance of poverty and sliding back down the social ladder into oblivion. In 1814, he could have given up the lights entirely, coasted on the success of the Bell Rock and built up a private business, as Telford and Rennie had done. He could have taken out patents on his optical inventions, looked for business that gave greater prestige or abandoned Scotland for more promising places. He did none of these. Work for the Commissioners at least guaranteed a regular salary. Work as a freelance consultant engineer, on the other hand, inevitably meant financial uncertainty. He took the path of caution, and paid for it in fame.

Robert was the most complex of men, a character who loved and courted physical fear but who was simultaneously terrified of emotional risk. In *Records of a Family of Engineers*, Louis noted Robert's 'interest in the whole page of experience, his perpetual quest and fine scent of all that seems romantic to a boy, his needless pomp of language, his excellent good sense, his unfeigned, unstained, unwearied human kindliness.' As Louis recognised, there were two competing forces at work in Robert, an unashamed pleasure in adventure, and a ruthless need for order.

> Perfection (with a capital P and violently underscored) was his design. A crack for a penknife, the waste of 'six-and-thirty shillings', 'the loss of a day or tide', in each of these he saw and was revolted by the finger of the sloven; and to spirits intense as his and immersed in vital undertakings, the slovenly is the dishonest, and wasted time is instantly translated into lives endangered. On this consistent idealism there is but one thing that now and then trenches with a touch of incongruity, and that is his love of the picturesque. As when he laid out a road on Hogarth's line of beauty; bade a foreman be careful, in quarrying, not 'to disfigure the island'; or regretted in a report that 'the great stone, called the Devil in the Hole, was blasted or broken down to make road-metal, and for other purposes of the work.'

Robert's perfectionism also manifested itself as an endless concern for money. The Commissioners' record books are littered with disputes over wages, requests for higher pay and demands for adequate pensions. Often, it was Robert arguing on behalf of others – for raising the pay of the keepers, or for helping the wives of those invalided out of the service. More usually, it was Robert worrying away at his own fear of being taken for granted. 'During the progress of the [lighthouse] work,' he wrote to the Commissioners in 1802 as part of a petition for an increase in salary, he 'travelled to the North sometimes by land and sometimes by water – ill provided with conveyance, exposed to many hardships and frequently in the greatest personal danger.' In 1808, when work on the Bell Rock loomed, the Commissioners had fixed his salary at £200 a year. In 1829, following a further demand, they doubled it. On his retirement, he was given a pension but was galled to discover that Alan had replaced him at a salary of £900 per annum. The remainder of his income came from private business, work for the Convention of Scottish Burghs and occasional articles in the learned journals. His expenditure was steady but burdensome. In addition to the cost of keeping a household, an office and a regular travelling itinerary, he was also devoted to the welfare of his apprentices. In part, he had arranged his life in order to allow himself freedom, but he had also deliberately chained himself to his duties and dependants. The more commitments he undertook, the more good works he piled up, the more firmly he was anchored to the bourgeois life. Security came from the immutable things in life – a family, a block of stone, a lighthouse, an annual salary. But all the fortune and stability in the world could not entirely silence his worries, and he ended up marking the same insecurities on his children.

The most significant of the relationships that Robert, and later Alan, forged with their engineering contemporaries was with a pair of Parisian brothers, Leonor and Augustin Fresnel. By the 1820s, the Fresnels were to optics what Smeaton was to sea-towers. Their work was singular for several reasons. Firstly, they were amongst the few scientists in Europe at the time to take the study of light seriously, and secondly, they devoted much of their time to finding practical applications for their experiments. Most importantly, from the Stevensons' point of view, the Fresnels were working closely on the perfect form of the lighthouse lens.

The link between the two families was first forged during Robert's tour of the French lights in 1820. Robert had taken himself off to see Corduan, the most famous of Europe's early lighthouses. The light had been built originally in the fifteenth century, and then repaired with much ceremony by the architect Louis de Foix in 1570 under orders from Henri III. Since the French King believed that all public works should reflect the greater glory of the monarchy, Corduan was reconstructed to look more like a classical temple than a lighthouse. Tiered like a wedding cake and decorated down to the last curlicue, it was the most elaborate lighthouse in the world, far exceeding even Winstanley's subsequent flourishes on the Eddystone. It took twenty-five years to complete and drove its architect almost to madness. Unfortunately, it was only after it was completed that de Foix realised the light was more exposed to the sea than he had initially supposed. In despair, he composed a poem to be engraved on the side of the building, challenging the gods to hurl their worst at his architectural wonder and cursing them for their indifference to his trials. In 1612, the gods responded. A bolt of lightning struck the top of the lighthouse and destroyed it. Construction work was taken over by another architect, Chatillon, who strengthened de Foix's foundations, while radical work in the 1780s swaddled the whole structure in a plain casing of fresh stone.

By the time that Robert saw it in the 1820s Corduan was more an example of archaeology than architecture, with so many succeeding layers of work outside and in that the original structure barely survived. It was less the building than the lighting that interested him, however. While Robert was still experimenting and refining his silvered reflector lamps in the Scottish lighthouses, Augustin Fresnel had taken a different approach and had sought to magnify the beam by placing reflectors behind the light and prisms in front. The difference between catoptric lights (in which the lens or reflector was placed behind the flame) and dioptric (in which the lens was placed in front) was akin to the difference between candles and electric light. Even the first tentative versions of the lenses strengthened the beam from a maximum of 1,000 candlepower to around 3,000 candle-power. Admittedly, the first lenses were clunky, primitive objects, more like giant myopic spectacles than the elegant prisms of later years. But the leap had been made, and all that remained was to refine the prototypes. Robert, having seen Augustin Fresnel's work, wrote to him and the two established an enthusiastic correspondence.

Fresnel's lenses also unwittingly became the subject of another of Robert's battles. When he returned to Scotland intending to test the effectiveness of lenses for use in the Northern Lights he discussed his findings with a friend, Dr David Brewster. Brewster, he knew, was interested in optics and prided himself on his knowledge of fashionable new sciences. But instead of being curious about Fresnel's works, Brewster was incensed. He claimed he had been the inventor of lenses for lighthouses, and produced as evidence a paper written some time previously advocating the use of 'polyzonal lenses' as 'burning instruments' for use in lights. He demanded that his lenses should be introduced to all the Scottish lights forthwith, and full credit given to him for their invention. Robert ignored Brewster's blusterings, but recommended to the Commissioners that lenses on the Fresnel model should be tested and then introduced gradually. A prototype was set up at Inchkeith light, and, though partially successful, needed modifications and was removed again for the time being. The Commissioners asked for further tests. Robert, being habitually cautious about major changes in designs and lighting, accepted their suggestions and began the slow process of adapting the lights for his purposes.

The NLB's prudence did not satisfy Brewster. In 1827, two years after the dispute had begun, he presented a paper to the Royal Society of Edinburgh 'On the theory and Construction of Polyzonal lenzes and their combination with Mirrors for the purpose of illumination in lighthouses'. In it, he accused the

Commissioners of pandering to 'sordid interests', and Robert himself of professional malpractice. Robert, furious, wrote to Brewster demanding a copy of the paper, and insisting that the 'mistakes' should be removed. Brewster refused. Robert wrote back, complaining that this was not the kind of behaviour expected of Royal Society members, and insisting that 'I have only again to state that I had no other object than to correct inaccuracies which seem to have crept into it.' Brewster again refused to let Robert see the paper, so Robert wrote to the RSE, demanding that the paper should be vetted by them and possibly withdrawn until the dispute had been settled. The RSE replied that Brewster's version should stand. Robert, reduced to impotent pique, announced that he 'decline[d] having anything further to do with the paper'.

It was soon evident that Brewster was not the kind of amiable opponent that Rennie had once been. He had a bilious temper and an infinite capacity to nurse a grudge. Once provoked, he considered himself engaged in a justified war for lenses over reflectors. In 1833, Robert finally tested the different equipment by setting up a display of French lenses, English lenses and Scots reflectors twelve miles from Edinburgh, which were viewed by the Commissioners from the top of Calton Hill. The Fresnel lenses, it was concluded, gave the strongest, steadiest beams, far superior to even the best of reflectors. Brewster, who appeared on the hill to press his case, wrote immediately to the Commissioners insisting that the results proved Robert's reactionary intentions and demanding that he, Brewster, should be appointed to the Northern Lighthouse Board to fit and supervise the new lenses. In a subsequent letter, his enthusiasm ran away with him and he declared that lenses should be fitted forthwith in every light in Scotland, that gas should be substituted for oil and, with a final flourish, that all the existing Scottish lighthouses should be 'dismantled' and rebuilt to suit the new specifications. The Commissioners, infuriated by Brewster's demands, delayed the results of their enquiries still further. Brewster, considering that the delays only proved the criminal intentions of the Board, enlisted Westminster in his cause.

Joseph Hume, a radical English MP, had made it a personal mission to investigate and remove all abuses of power within the public service. It was a laudable aim with meddlesome ends. The Scottish lights worked well as they were and Hume had little to teach the Commissioners about either parsimony or bureaucracy. Nevertheless, Hume established a Parliamentary Select Committee of forty-six members in 1834, ordering them to conduct a thorough review of all the lighthouse authorities and to investigate Brewster's allegations.

The Committee studied everything from the way the separate administrations were managed to the relative benefits of different fuels. The NLB Secretary Charles Cuningham, the Commissioner James Maconochie, Robert and Alan Stevenson were all called before the Members to vouch for the NLB's works. Between them they answered over 1,200 questions on the Scottish lights. The Committee's report recommended sweeping changes to both the English and the Scottish services. Trinity House, the Irish lights and the NLB were to be merged and given a base in London; light dues were to be paid to the Treasury who would then parcel out a sum to the centralised administration, and the English lights in private ownership were to be nationalised. In Scotland's case, the Committee also proposed that all work on the lights should be done by locals, and hinted darkly that they considered the NLB's chief engineer had altogether too much power.

Uproar ensued. An anonymous letter to the *Edinburgh Evening Post*, probably written by Captain Wemyss, one of the Commissioners, pounded out its disapproval of both the report and the subsequent bill. 'The motion,' said the writer, 'is nothing more or less than a barefaced, unpatriotic and absurd proposition to place the Lights of Scotland under the Trinity Board in London . . . Let Scotland take alarm at such bold inroads on her national individuality and let her point the finger of contempt and raise the slogan of shame on her Humes, Murrays, Fergusons and such representatives . . . In conclusion, I would merely enquire whether that algebraical pated blockhead Hume knows anything at all of the subject whereon he prates so glibly?' Robert himself ridiculed both the suggestion that he would be able to find well-trained masons in the isolated settlements of the isles and the plan for amalgamating the three administrations. Trinity House, he pointed out, had been criticised in the report for 'jobbing and plunder', its lighthouses were a disgrace, and the service was still being run for individual profit, not public benefit. The Scottish lights, by comparison, were prudently managed, successful, and relied almost entirely on close local understanding. Why then should the inefficient English service take over the efficient Scots one? The Irish service made similar objections and the bill was acrimoniously debated in the Commons.

Hume finally backed down a little. The three administrations would be allowed to remain separate, he conceded, but Trinity House was nevertheless to take on an overall supervisory role. All new lighthouse work had to be authorised by them and no new lighthouses could be built until Trinity House allowed it. Separate tolls would be abolished for the Scottish and Irish lights, and there would now be one due paid by all shipping round the British coast. The compromise satisfied no one. Brewster had found no better support for his cause in London than he had in Edinburgh, and as far as the Commissioners were concerned, the Act, even in modified form, interfered with their independence. The whole process had taken months of argument, time that would have been better spent on practicalities. Brewster, never one to accept defeat with good grace, did not diminish his campaign. Indeed, time and parliamentary indifference only seemed to make him angrier.

Though Fresnel's lenses were gradually accepted and adopted in all the Scottish lights – following a much more thorough enquiry by Alan Stevenson – Brewster still brooded. Almost thirty-five years after the original dispute had begun, Brewster continued hurling broadsides at Robert and all his descendants. He wrote a piece for the *Scotsman* in June 1860 announcing 'a Review of the Conduct and Writings of Messrs Robert, Alan, David and Thomas Stevenson, as engineers to the Scottish Lighthouse Board, in reference to their Ignorance of the proper optical arrangements for Lighthouses and Distinguishing Lights – their perversion of Scientific History – their interested and obstinate opposition to the substitution of Lenses for hammered Reflectors – and their calumnies against the inventors of the Dioptric System of Lights, now in universal use, introduced into Great Britain by the persevering labours of Sir David Brewster and into France by the celebrated philosopher M A Fresnel.' The audience for this dispute had clearly grown exhausted some time ago. The *Scotsman* itself warned 'against any third party attempting to act as judge or umpire between the combatants', and a letter to Brewster from M. Biot of the French Academy of Science announced peremptorily that 'he conceived the wrongs Sir D Brewster complained of to be purely imaginary, and concluded that by saying that at their time of life such retrospective polemics should be avoided.'

The whole dispute had achieved almost nothing except proof of Brewster's temper and Robert's over-cautious habits. Perhaps some of the public humiliations, and much of the parliamentary wranglings, could have been avoided if Robert had been keener to adopt new technology; perhaps there was a speck of truth in Brewster's allegation that the Commissioners' chief engineer was granted too much power. But the core of the argument – that Brewster, not Fresnel, should be credited with the invention of lighthouse lenses and that the Commissioners and Robert had acted criminally in failing to adopt them – had been proved false. The only useful consequence of the saga was to demonstrate beyond doubt that Robert's successors were more than worthy of their name and position. Robert's dithering over the adoption of lenses and his habit of retreating into bluster and pomp when threatened seemed at odds with Alan's intelligent analysis of the new methods. It had been Alan who had understood the science of lenses, Alan who had adapted them for Scottish purposes, and Alan who had done most to ensure their success. Robert might have made more noise, but Alan, it seemed, was going to be more than a match for his father.

SIX

Skerryvore

The helicopter skims low over Mull, dipping past swatches of forestry and fox-red moorland. Out beyond, past Staffa and Fingal's Cave, lies the Atlantic, placid today but easily roused. Away to the left juts the remnant edges of the Scottish mainland; up on the right, Tiree and Coll appear flattened against the horizon. The helicopter flies on, low over the water, past ladders of sunlight and clusters of rock.

Finally, just when the passengers can see nothing but the width of the ocean and the size of the sky, there is a flash of whiteness up ahead. At first it's only a disturbance in the water, then a small blackened stub appears, rising up out of a ruff of surf. A little closer, and the passengers can see a tangle of black rocks stretching away to the left with the sea beating itself repeatedly against their sides. Rising up from the centre of the reef, like the spire of some subterranean cathedral, is a dark tower. On its crown is a diamond-paned lantern, a weather vane and a balcony rail; down the sides are tiny slitted windows like a row of buttons. To one side is a rudimentary pier, and on the right a concrete pad marked with an 'H'. A few whiskery seals watch the helicopter's approach, then flump off the rocks into the water to join the sea birds. The helicopter lands and the passengers scurry away to crouch under the lee of the tower.

Up at the top of a precipitous iron ladder and through the nine-foot thickness of granite, there is a metal door, barred and padlocked. Inside, there are more ladders, a confusion of machinery and a strong smell of neglect. The rooms of the tower reach up and up, through an endless succession of batteries, generators and flickering technology. The only break is for a tiny kitchen (as cosily fitted as anyone could wish) and two cramped little rooms, containing narrow bunk beds and a port-holed window. Up and up, one ascends past more machinery, more ladders, more clutter, and finally to the light room. Outside, the wind thuds against the walls. From the balcony rail, there is a sudden overwhelming landscape of faraway islands and ocean. At your feet lie hundreds of dead birds, guillemots and gulls, blackbirds and curlews. During the migrating season, the lantern becomes an immense candle courted by giant moths. The birds flock in such huge numbers here that it is considered too dangerous to go out on the balcony. Up above, past the cranes and aerials, is the diamond-patterned lantern. Inside, three circular lenses revolve silently round the light, catching and refracting the weak daylight so the bulb appears by turns large and small, large and small. A cardboard box on the floor contains a few replacements, each the size of a punctured rugby ball.

A fire during the 1950s gutted much of Skerryvore. Automation took the rest. All that remains of its creator is a few wrought-iron sea serpents holding up a railing in the lantern, and the tower itself. That tower is still extraordinary. Walk slowly around the curve of the base, and it looks for all the world as if it grew from the rocks of its own accord. The dips and summits of the reef fit the walls so closely that it is difficult to work out which parts are nature and which artifice. The first few courses are black Tiree gneiss, as organic as the roots of an old tree. Further up, the stone is pinkish. From a distance it looks like the last surviving remnant of a petrified forest. Skerryvore has been described as the most beautiful lighthouse in the world. It is twelve miles from the nearest land, and was built to be avoided.

If you take the more conventional route to Skerryvore, you catch a different angle. The daily CalMac ferry from Oban to

Tiree passes round the Sound of Mull, stops at Coll and then docks finally near a broad sweep of honey-coloured beach. Tiree is a treeless island, beaten almost flat by the winds hurtling overhead on their way to the mainland. At the far end of Gott beach, beyond the bright insect wings of the wind-surfers, is a ramp of black rocks and a few scattered houses. The shore is thick with flotsam – plastic bottles, old nylon rope, fishboxes, margarine tubs. Buried in the coarse grass every few yards is a rusty iron buoy, solid as a bomb. At the other end of the island, past the holiday-cottage blackhouses with their humpbacked roofs, is Hynish. There's a picturesque cove with a pristine harbour, several sturdy workmen's cottages, recently restored and now used for Outward Bound courses, and a lacework of stone walls, crumbling in places but still intact. At the top of a small hummock is a squat tower shaped like the butt-end of a Victorian castle. Inside the staircase winds upwards past a succession of fading posters and photographs – illustrations of building works, pictures of bowler-hatted workmen, a *Scotsman* obituary to an engineer. An immense bell lies in a corner stamped with the imprint of a lighthouse and the words '*In Salutem Omnium*'. At the top are a collection of museum pieces, a silvery argand lamp with its wick still unlit, a storm lantern and a pair of binoculars fixed against the wall gazing blindly out towards the sea. The place looks deserted save for the lone staggering inhabitant of the keepers' cottages walking his dog round the houses again and again.

A little beyond there is a beach known incongruously as Happy Valley. The rocks which surround it have been rolled by the sea into fantastical shapes, such as chairs, shelves, pools, secret hiding places. It is a very different place from the cur-vaceous surfers' haven on the other side of the island. There is no sand here and no flotsam, only the sea and the wind. Up above the beach is a black promontory of rock scattered with fragments of seashell and grass nibbled close by the ubiquitous sheep. On a mild day in early October, it is still too windy to light a cigarette. From the top, if you squint out towards the watery horizon, you can see a flash of spray and a tiny grey spike. It seems barely there; it is the white water below it rather than the thin pencilled line that draws your attention. Beyond it there is nothing at all.

Two hundred years ago, when the stones in Happy Valley were a little less smooth, that distant stub brought its rewards. Skerryvore wrecked ships year after year, and every time it did so, the fragments – wood, cargo, broken bodies – would drift towards Tiree and fetch up on one of the nearby beaches. By the 1830s, the rewards from the reef were considered so reliable that, as Robert had found on Sanday, rents on the Hynish side of the island remained higher than elsewhere. The east side stayed poor; the west side got rich on a steady harvest of wreck. When the NLB's clerk of works came to assess the damage done, he calculated that at least thirty ships had been destroyed on Skerryvore between 1804 and 1844. He drew up a list, though, as he pointed out, it was not comprehensive. 'Very many vessels were wrecked on this dangerous reef whose names could never be learned,' he wrote later, 'and of which nothing but portions of the drift wood or cargo came ashore; and there have, no doubt, been many shipwrecks of which not a single trace has been left. Nothing, indeed, is more probable than that many of the foreign vessels whose course lay through the North Irish Channel, and whose fate has been briefly and vaguely described, as "foundered at sea", have met their fate on the *infames scopuli* of the Skerryvore.' The Tiree fishermen, he noted, were in the 'constant practice' of sailing out to Skerryvore after a storm in the hope of finding wreck trapped in the rocks. Time and the lighthouse have evened out conditions on the island, and now there are few indications of prosperity in Happy Valley.

By the 1830s, Skerryvore had been bothering the consciences of the Commissioners of Northern Lights for over forty years. Letters pleading for a light had been arriving in George Street almost on a weekly basis. Suggested designs for towers made of stone, cast iron, even bronze were submitted by impatient amateurs. By 1835, when the Board invited interested parties to submit their opinions, the stream broadened to a flood. The Commanders of Revenue Cruisers, the Committee of the Glasgow Chambers of Commerce and the Chairman of the Liverpool Ship Owners Association all pressed the point. 'I have frequently passed the rocks of Skerrivore,' wrote the Inspector General of the Leith Coastguard, Captain Knight, 'and consider them of so dangerous a nature and so completely in the direct Track of Vessels that I have no doubt many are wrecked on them and never heard of.' James Melville, captain of the Revenue Cruiser *Swift*, agreed: 'I am fully satisfied that there is not a station on that Coast, where a light is required more urgently for the safety of Vessels than upon these dangerous Rocks.' The Board recorded only the petitions of the governmental organisations and the shipowners. The sailors who were most at risk from Skerryvore – the islanders and the local fishermen – stayed silent. Money and politics, as usual, shouted loudest, but there was also a darker undercurrent to the islanders' silence. In part, it was the old suspicion of meddling southern men and the lure of profits from wreck. In part, it was still the settled belief that there was no need for a light on the Skerryvore rocks; God had placed those rocks there, God had meant them to be a warning to the unwary, and if God had meant a lighthouse to be built on the rocks, he would have put it there himself. Sailors and islanders are traditionally fatalistic. In the case of Skerryvore they had good reason to be.

Robert Stevenson had visited the reef twice: once in 1804, and once during his pleasure-tripping inspection tour with Sir Walter Scott in 1814. Scott's glum assessment of the place ('a most desolate position for a lighthouse – the Bell Rock and Eddystone a joke to it') had not changed in the intervening years. Occasionally, the subject would reappear in NLB meetings and then would be quietly shelved again. Robert had never doubted the need for a light on Skerryvore. The Commissioners, ensconced in Edinburgh and surrounded by fearful accounts of Skerryvore's reputation, were more cautious. The expense was ridiculous, they argued. They had projects of equal urgency to consider and they were still unsatisfied that a light was fully feasible. Finally, in 1834, they sent Robert back to Skerryvore, this time to survey the rock in more detail and to report on possible sites for a light. Robert set off for the isles, dragging Tom and Alan in tow.

On his return to George Street, he presented his report:

> The Rocks of Skerrivore lie 12 miles South of the Island of Tyrii and about 35 Miles from either of the Light Houses of Barrahead and Rhins of Islay. This Reef forms a very great Bar to the Navigation of the outer passage of the Highlands, a Track which is used chiefly by His Majesty's Vessels of War and first Class of Merchantmen. It is also an obstacle of no small magnitude to the foreign Trade of the Clyde and Mersey. The Rock on which it is proposed to build this Light House forms the foreland of an extensive track of foul ground lying off the Coast of Argyllshire. This reef has long been the terror of the Mariner, but the Erection of a Light House upon Skerrivore would at once change its Character and render it a rallying point of the Shipping which frequents these seas.

The reef, he concluded, ran for seven miles in total, much of which was permanently submerged underwater. Skerryvore itself was only the highest of a cluster of rocks: Boinshly (the 'deceitful bottom') lay around three and a half miles away from the proposed site; Bo-rhua (the 'red rock') lay between Boinshly and the main reef.

Skerryvore ran for around 280 feet in total, much of which remained underwater even at low tide. Between each of the major rocks flowed narrow channels of water that looked tempt-

ing enough for small-boat sailors but were primed with underwater hazards to trap the unwary. None of the three separate
reefs made an easy site on which to build. The main reef was
pocked with peaks and gullies and surrounded with a rampart
of detached rocks. Much of the surface was so irregular in places
that the movement of the water flung spouts of water on to the
reef like the blowholes of whales. And on those parts of the
reef that took the full weight of the sea, the rocks had been
worn smooth, almost icy. Even more awkwardly, the sea had
eroded much of the reef's underwater surface, gashing it with
deep subterranean trenches under apparently solid gneiss. As a
site for a lighthouse, Skerryvore could hardly have been
worse.

Robert, however, seemed confident of his chances. At the end
of his report, he blithely claimed that 'the Erection of a Light
House on the Chief of the Skerrivore rocks is not only a practicable work, but one which will be much less difficult and expensive than that of the Bell Rock.' It was a strange judgement.
Robert, of all people, understood the hazards of building tower
lights and knew not to underestimate their complexities. Admittedly, unlike the Bell Rock, some part of Skerryvore was always
out of the water even at high tide. But in every other respect,
Skerryvore was an altogether more daunting proposition. The
Bell Rock was at least one solid reef, whereas Skerryvore lay
scattered like miniature mountains over many miles of sea. The
Bell Rock was also sheltered a little by the bulk of the east coast,
while Skerryvore was stuck right in the path of every Atlantic
gale roaring over the seas from Newfoundland. Those storms,
as Robert well knew, could reach a pitch of ferocity and destruc-
tiveness that the North Sea rarely matched. More seriously, the
area surrounding the Bell Rock was well known and well surveyed. The waters around Skerryvore, by contrast, lacked even
the most basic soundings. Admiralty charts of the west coast,
the Minch and the Hebridean Sea still contain warnings that
much of the area has not been surveyed since 1856. As Alan

noted, they had to begin from scratch with Skerryvore. During the preliminary surveys of 1834,

> several vessels came so near the Rocks as to cause, in the minds of the surveyors, who witnessed their temerity, serious fears for their safety. On one occasion in particular, a large vessel belonging to Yarmouth, with a cargo of timber, was actually boarded between Mackenzie's Rock and the main Rock of Skerryvore by the surveyors, who warned the Master of his danger in having so nearly approached these rocks, of the existence of which his chart gave no indication. On another occasion, a vessel belonging to Newcastle was boarded while passing between Bo-Rhua and the main Rock; and so little indeed had the Master (whose chart terminated with the main rock and shewed nothing of Bo-Rhua) been dreaming of danger, or fancying that he was in a cable's length of the reef, that he was found lying at ease on the companion, enjoying his pipe, with his wife beside him knitting stockings.

Perhaps the worst hazard of all at Skerryvore was the lack of any adequate shelter. The Bell Rock works had been well equipped with all necessary materials – granite from Aberdeen and Edinburgh, boats and supplies from Leith and a well-trained workforce drawn from the many fishing villages along the east coast. Skerryvore only had Tiree. From an engineer's point of view, the island was a dismal prospect. It had no harbour and no shelter for shipping. It had no raw materials. Everything, from stone to wood to workers, had to be imported. The land was sandy, and the trees were brought by boat. 'It is said,' noted Alan, 'that this total absence of fuel in Tyree is the result of the reckless manner in which it was wasted in former days in the preparation of whisky; but, however this may be, certain it is that the want of fuel greatly depresses the condition of the people.' If the NLB seriously intended to build a light near the island, then 'craftsmen of every sort were to be transported, houses were to be built for their reception, provisions and fuel were to be imported, and tools and implements of every kind were to be made.'

Perhaps Robert's belief that a light on Skerryvore would be a simple business was merely a piece of shrewd politicking, designed to lull the Commissioners into advancing the money. If so, it was not a theory that worked. They peered again at Robert's estimate (£63,000), then at their balance books, and then at the glut of building works already underway. They set up a special Skerryvore Committee to deliberate further. Finally, they insisted on making a journey to Tiree in the summer of 1835 to see for themselves. During the trip, their steamer lost its boiler rivets just off Skerryvore itself, and a fire broke out in the boiler room, crippling the ship. Though the fire was eventually extinguished, the experience concentrated the minds of the Commissioners wonderfully. Being stuck on a disabled ship just off the most dangerous rocks in Scotland was, they decided, a remarkably persuasive argument for a light on the Skerryvore, no matter what the expense.

The person elected as resident engineer for the works, however, was not Robert, it was Alan. By 1835, Robert Stevenson was sixty-three, still as energetic as ever, but clearly too old to spend four years heaving stones around on a sea-drenched reef in mid-ocean. In the twenty-four years that had passed since he completed the Bell Rock his business had swelled, his standing increased and his children had grown up. The constant journeying had kept Robert fit. Time and fine living had given him substance without fat. In portraits, he had kept the fierce gaze of his youth, but gained a weathered glow to his cheeks. Somewhere in his expression there remained that strange deceptive twinkle of his youth. He felt no particular weakness and remained as sharp as ever in mind and limb. But, as he grudgingly acknowledged, he was in no position to attempt such an immense and potentially debilitating project. His professional aim over the last thirty years had been to build himself a reputation as a technical expert, to be an authority, not a builder's mate. From the mid-1820s onwards, he had taken a loftier role both in private business and with the lighthouses. Having made his recommendations, he would move on, leaving a trusted band of Stevensonians to complete the work on his behalf. There was no longer any need for Robert to paddle through storms or quarrel over dovetails; there were others to do it for him. Besides, he had spent the last twelve years patiently training up his eldest son to take on exactly this kind of task.

By 1835, Alan Stevenson had been an apprentice for seven years, Clerk of Works to the Northern Lighthouses for a further five and an equal partner to his father in the family business for two. At the age of twenty-eight, he had already designed and built seven lighthouses under Robert's supervision, and had recently undertaken a complete review of all lens designs in the Scottish lights. His path to full qualification had been a weary one, dulled with bureaucracy and routine. But if his father remained hesitant about handing over full powers to Alan, the Commissioners didn't. They trusted him, appreciated his quiet, incisive intellect, and believed in his future. Skerryvore, it was evident, was the perfect moment to launch Alan as an architect and engineer in his own right.

Robert's feelings towards his eldest son remained ambivalent. It was plain to Robert, as it was to anyone else who came in contact with him, that Alan was an exceptional young man, more than capable of shouldering the full burden of the Stevenson engineering business. But Robert veered between pride and incomprehension. He was well aware that Alan was clever; too clever, Robert considered, for his own good. Robert had never fully recovered from Alan's early flash of independence and still fretted when he heard reports of Alan's literary leanings or came across scribbled fragments of poetry. In 1828, alerted to a further bundle of verses, Robert wrote worriedly to Alan, reminding him that 'This is a very precious time for you, Alan, in the study of elementary and technical books till you can no more forget them than that you have ten fingers and as many toes. It is a great pity that we too often let such opportunities slip. Yet surely there is nothing in your present circumstances that should distract your attention from the theory and practice of your profession.' Alan was then twenty-one, had been an apprentice engineer for six years and showed no more inclination to take up a career in literature than he did for knitting. He loved the world and its sensations for their own sake, not as a substitute for his profession. But his enthusiasms bothered his father deeply. All the extraneous stuff of Alan's life – the books, the poems, the obscure classical allusions he sometimes inserted into professional reports, the love of travelling and the soft-hearted thoughtfulness – were puzzling to Robert and at times downright suspect.

The travel, in particular, worried Robert, and he occasionally vented his frustrations to others. While Alan was in France in 1824 working with the Fresnel brothers and touring engineering projects, Robert wrote to his assistant Alexander Slight, complaining that his son would have been more use at home in Scotland working on the new light at the Rinns of Islay than frittering away his time in foreign places: 'His writing to you of his travels will naturally draw upon you the obligation of your opinion as to the amendment to be made by application and a steady pursuit of the immediate object in which he is engaged.' If Robert alone couldn't persuade Alan to sit down and behave as he should, in other words, he was going to have to recruit others to help concentrate his mind. No matter that Alan had taken the trip with Robert's encouragement in the first place. While he remained out of sight, there was no telling what distractions he might find. Robert knew his son to be a talented and conscientious worker, but could not bring himself to trust him. So he imposed absurdly high standards on Alan, forcing him onwards to an endless succession of false summits, testing his loyalties again and again. Somewhere at the back of Robert's exhortations, it is possible to detect a small note of fear.

Despite all his efforts to do his father's bidding, Alan could never entirely dampen his pleasure in the outside world. Travel brought out his romantic habits. Away from the suppressions of Edinburgh, he developed a sharp eye and a talent for details. Much of his time abroad was spent working, touring lighthouses in France, canals and bridges in Stockholm or roads in Moscow on the instruction of Robert or the Commissioners. While he remained away from home, Alan knew that Robert would be sitting irritably at home in Baxter Street, fussing over his studies and fearing for his morals. Confined within his father's expectations so much of the time, he did what most clever children do and turned himself into an astute psychologist. He learned to speak Robert's language and realised that the best way of pacifying him was to anticipate his worries. He would write home often, giving descriptions of the works he had seen and the people he had met and inserting the odd reassuring allusion to Robert's works. He compared the superiority of Scottish engineering to the pathetic specimens he found abroad, made disparaging mention of Popish practices and usually concluded with a homesick joke or two. He was usually careful to include a brief account of his attendance at church, and to pick up on a couple of points from the sermon. If he was tempted to quote poetry, he confined it to short bursts of Sir Walter Scott. Sometime in 1827, he changed abruptly from signing himself, 'your most loving son', to 'your most dutiful son'.

Aside from feeding his restless mind, travel addressed something more straightforward. Alan was, and always would be, prone to sickliness. As for his nephew Louis many years later, time spent abroad was also time spent chasing health. Travel was Alan's self-administered medicine. His letters from abroad rarely mention illness, but at home in Scotland his work was blighted by intervals confined to bed. It was not an argument

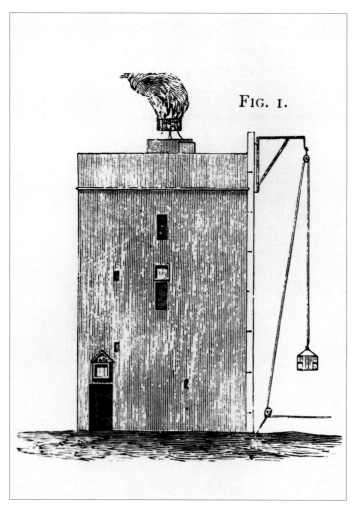

FIG. I.

The Isle of May, Scotland's first lighthouse, was built in 1636; it featured a coal-fired brazier on top and a winch for hauling up fuel.

Winstanley's Eddystone lighthouse, completed in 1699 and destroyed in 1703 by a storm in which Winstanley himself was drowned.

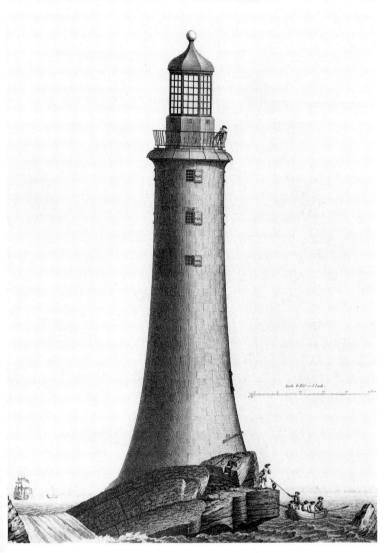

Smeaton's Eddystone lighthouse, with its dovetailed stone courses, was completed in 1759 and became a model for further lighthouse construction.

John Rennie's design for the Bell Rock shows similarities to Smeaton's Eddystone design.

Fig. 15.

Below Robert Stevenson's design for the Bell Rock light, completed in 1811.

Right Cross-section of the Bell Rock.

Robert Stevenson, head of the Stevenson engineering dynasty, 1814.

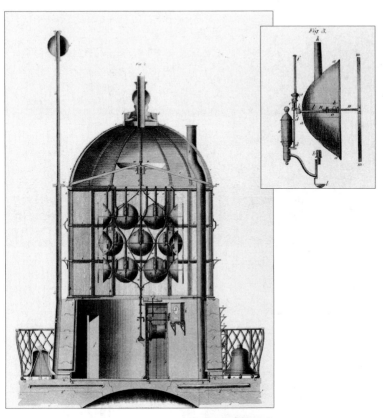

Above The Bell Rock light room, with its system of reflectors and winding mechanism.

Right Robert Stevenson in later life, 'a man of the most zealous industry, greedy of occupation, greedy of knowledge . . . unflagging in his task of self-improvement'.

J.M.W. Turner's illustration of the Bell Rock during a storm, commissioned by Robert Stevenson.

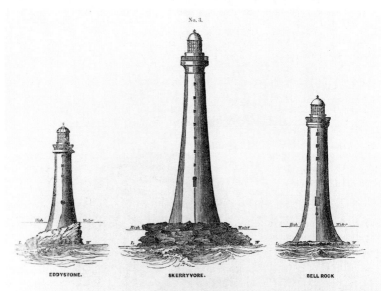

No. 3.

EDDYSTONE.   SKERRYVORE.   BELL ROCK

Lighthouse	Height of Tower above first entire course in feet (H)	Diameter in feet		Distance of centre of gravity in feet from base. (G)
		at Base	at Top	
Eddystone	68	26	15	15.92
Bell Rock	100	42	15	23.59
Skerryvore	138.5	42	16	34.95

Alan Stevenson's Skerryvore light, completed in 1844.

Inset The only known portrait of Alan Stevenson.

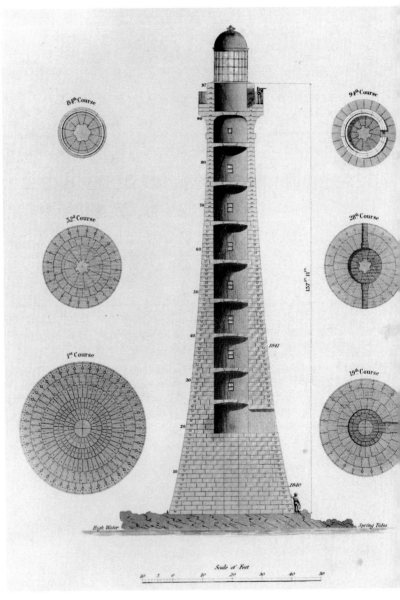

Cross-section of Skerryvore, showing details of the 97 stonework courses.

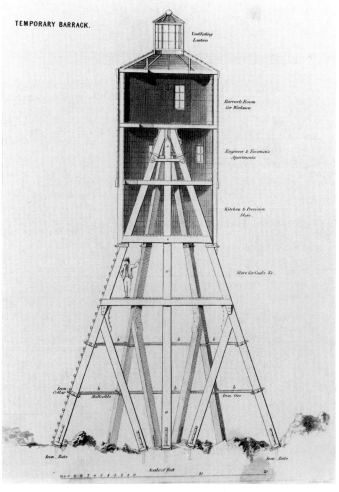

TEMPORARY BARRACK.

Ventilating Lantern

Barrack Room for Workmen

Engineer & Foreman's Apartments

Kitchen & Provision Store

Store for Goods &c.

Iron Collar

Malleable

Iron ties

Iron Bats

Iron Bats

Scale of Feet

The temporary barracks at Skerryvore, rebuilt after the original structure was destroyed by gales.

Alan's design for Ardnamurchan, Scotland's only Egyptian-style lighthouse.

Above The impossible lighthouse, Muckle Flugga, built in 1854 by David Stevenson at Scotland's most northerly point to guide British naval convoys on their way to the Crimea.

Right David Stevenson, 1815–1886.

Left Thomas Stevenson, 'a man of somewhat antique strain . . . shrewd and childish, passionately attached, passionately prejudiced; a man of many extremes'.

Below Robert Louis Stevenson, the reluctant engineer.

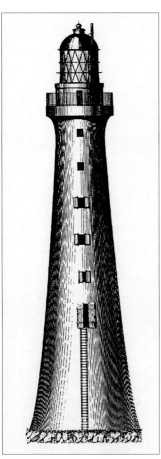

Above The light of Dhu Heartach, built under the supervision of David, Thomas and Robert Louis Stevenson, and completed in 1872.

Right Raising coal to the tower at Dhu Heartach.

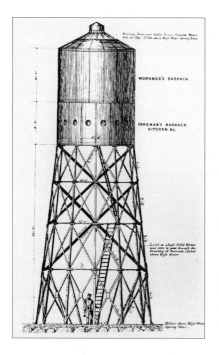

Drawing labels:
During Summer Gales loose broken Water fell on Top. 77 feet above High Water Spring Tides.

WORKMEN'S BARRACK.

FOREMAN'S BARRACK. KITCHEN &C.

Level at which Solid Water was seen to pour through the Framing of Barrack 35feet above High Water.

30 Feet above High Water Spring Tides.

Left The barracks on Dhu Heartach, in which the resident engineer, Alan Brebner, and thirteen work-men were trapped for two weeks during a gale.

Below Dhu Heartach (or Dubh Artach, as it is now). Lightkeepers disliked it for being small, cramped and uncomfortable. Like all of the Scottish lights, it has now been automated.

that would have persuaded Robert, least of all when Alan's chosen cure seemed so luxurious. But Robert could scarcely have complained that Alan spent his time travelling in languid convalescence. When the Commissioners sent him to France in 1834 to research lenses with Leonor Fresnel, he returned with results that changed Scotland's lighthouse optics for good. He had already met the Fresnel brothers in 1824, and the three had established a productive professional friendship.

During the complex transition from reflectors to lenses in the Scottish lights, Alan corresponded with the brothers frequently, keeping in touch with their experiments and encouraging their practical applications. He also escorted Leonor and his wife on their tour of Scotland in 1837, visiting the lights and inspecting Stevenson works. 'Fresnel is in high spirits,' wrote Alan from Thurso in August, 'he and Madame sung a scene of an opera after breakfast in capital style.' Alan, the Fresnels discovered, provided a willing audience for their work. He was lively, quick-minded and less suspicious of new technologies than his father. By the late 1820s he knew the Scottish lights well enough to decide which designs would or would not work within them and suggested refinements, first with diffidence and then with gathering self-belief. When Augustin Fresnel died in 1828, Alan continued his association with Leonor, who introduced him to his lens-maker, Soleil. 'After seeing the lens apparatus in its present improved state I confess I think it possesses some notable advantages over the reflecting apparatus,' he wrote home to Robert in 1834, 'and what I least of all expected is the apparent superior fitness for Fixed lights ... On the subject of immediately dismantling our present lights, to substitute lenses, M. Fresnel thinks our Reflectors so much better in England and Scotland than those of France that there is not the same call for immediately adopting lenses ... Wait, he says, until your present reflectors are worn out and then replace them by lenses.' When Alan did begin establishing a full set of dioptric lenses in the Scottish lights, it was, by default or design, Robert who took much of the credit for their development. Alan wisely stayed silent and waited his turn.

At home, working for the Stevenson firm or the Northern Lights, Alan seemed everything his father wanted him to be. He put up with the long hours of transcribing reports, preparing specifications and plotting budgets without complaint. He attended to the details of less glamorous projects with efficiency and patience. Despite his occasional ill health, he spent the same days flogging from place to place that Robert and Thomas had in their time, sometimes confined to his cabin with seasickness or stuck in damp lodgings far from the nearest town. 'I have taken your advice about my flannel shift, which I found repeated on the outside of the letter, no doubt on my mother's advice,' he wrote to his father while surveying sites for a railway in Forfarshire. 'Such husbanding of health, it must be said, for one who has walked wet and dry through Strathmore!' Finally, in 1830, seven years after he had first begun work for Robert, the Commissioners of Northern Lighthouses appointed Alan their Clerk of Works.

The appointment was made on Robert's recommendation, and caused some controversy at the time since Robert had already promised the post to at least one other of his assistants. In a letter to James Slight as far back as 1823, he had discussed his plans for the position, and then confessed, 'I would not have thought that I did right if I had not put this offer in your way.' Between that time and Alan's appointment, the post had remained open. Perhaps Robert had always intended that his son should fill it; perhaps, judging by his political thinking, he had offered it to his other assistants simply as a method of encouraging them to be more competitive. Whatever the reason, it was evident that Alan had paid his dues to the lights many times over. As he noted in his official address to the Commissioners, 'I am therefore inclined to hope that I have enjoyed every opportunity of becoming professionally acquainted with the Engineers department of the Light House

Service and I take the liberty of expressing my earnest desire
in the event of my appointment to the situation of the Clerk
of Works to fulfil with fidelity the trust committed to me.'
The Board appointed him unanimously on a salary of £150 per
annum. His workload, already heavy, increased immediately.
Six new lighthouses had to be planned, designed and built within
the space of three years. Each one was scattered as far from the
others as geographically possible. Somehow, through perseverance and efficiency, all the lights were finished on time.

And then in 1834 came Skerryvore. As public works, any
new plan for the lighthouses had to lumber through a lengthy
bureaucratic process before building could begin. The Commissioners had already gained parliamentary authority for the
light, but they needed to have the construction money advanced,
buy sufficient land from Tiree's landlord, the Duke of Argyll,
for a workyard and houses, and now, after the intervention of
Sir Joseph Hume, have their plans vetted by the guardians of
English lights. In September 1836, empowered by their new
role as supervisors of the Scottish lights, a deputation from
Trinity House made a trip out to the reef. The sea was pearly-
smooth, the weather was fine and the Elder Brethren landed
on the rock without so much as wetting their epaulettes. 'A
lighthouse no doubt can be built here without difficulty,' the
Brethren noted unconcernedly in their report, 'as it is sufficiently large and high for the workmen to remain on the rock
. . . The Commissioners are making preparation for the erection
of a lighthouse on the largest Skerryvore, and it is probable that
ere long the Court will be called on for an opinion on the
subject.' Finally, four years on from Robert's survey, after the
Elder Brethren had departed, after the Duke of Argyll had been
petitioned for fifteen acres, and after the Skerryvore Committee
had picked over the finances to its satisfaction, building was
authorised to commence. On 22 February 1838, Robert Bruce,
the Sheriff of Argyllshire, proposed to the Committee that Alan
Stevenson, Clerk of Works, 'was the most proper person to be

employed as resident Engineer for the important works on the Rocks.' His salary, accordingly, was to be supplemented with an extra six guineas a week forthwith.

During the summers of 1836 and 1837 Alan had been busy. As clerk of works he was responsible for overseeing the preliminary arrangements for the light, searching for a suitable quarry, purchasing land, constructing a harbour, hiring workmen, transporting materials, setting up houses and ensuring that the works were well stocked with the best materials. Alan reluctantly chose the bay of Hynish as the most suitable spot for the work-yard on the basis that it was the closest land to Skerryvore. He had few illusions about the site. The whole island of Tiree, he considered, was hopelessly exposed to the wind and the swell, and Hynish, however picturesque, was no better than the rest. But it was the only land from which the reef could be seen. He had no choice but to build there. 'I know by too frequent experience,' he acknowledged later, 'that the embarkation in the Bay is a matter of great difficulty and hazard.' One hundred and sixty years later, the islanders are still sour about his choice. The bay where Alan built his dock and sluice system, they point out, is prone to such bad swells that it is impossible to moor there on all but the calmest summer days. The Commissioners' response was to point out acidly that Hynish was not built for the greater good of the island, but to service the lighthouse. Even so, it was a contrary choice. The dock still stands, as fine a piece of engineering as any Stevenson ever contrived, but, on an island without any decent harbour, it is almost entirely useless.

Alan also had workmen to hire. Thirty trained masons, twelve quarriers, four smiths, two foremen and a shoal of contract carpenters, builders, joiners and storekeepers were needed. Each was to be paid between 3s 10d and 2s 6d per day for two years' guaranteed work. Alan was in charge of the health and welfare of every single one. He took only a few of the labourers from

Tiree itself. He did occasionally use local boatmen and quarriers but found that 'partly from incapacity and partly from excessive indolence, [they] could not be trusted for a moment to themselves.' The islanders were suspicious of the Edinburgh men; the workmen, in turn, treated the islanders with contempt. Even Alan was not above the occasional twinge of southern smugness. Once the workyard was up and running, he noted.

> the desolation and misery of the surrounding hamlets of Tyree seemed to enhance the satisfaction of looking on our small colony, where about 150 souls were collected in a neat quadrangle of cleanly houses, conspicuous by their chimnies and windows amongst the hovels of the poor Hebrideans ... The regular meals and comfortable lodgings and the cleanly and energetic habits of the Lowland workmen, whose days were spent in toil and their evenings, most generally, in the sober recreations of reading and singing, formed a cheering contrast to the listless, dispirited, and squalid look of the poor Celts, who have none of the comforts of civilised life and are equally ignorant of the values of time and the pleasures of activity.

Alan neglected to mention that Tiree had just been scourged by the Black Factor, the Duke of Argyll's lieutenant during the Clearances, who was to reduce the population from over 4,400 in 1831 to 3,200 in 1861. It was not entirely surprising that the islanders seemed depressed when most had lost a family member to the emigrant ships.

Alan also began looking for suitable stone for the light and instructed one of his assistants to begin quarrying the black granite on Tiree. In the space of two summers, they turned out 3,800 cubic feet of rock, which was eventually used for the lowest four courses of the light. But the Tiree stone was too flawed, too hard and too time-consuming to use on the whole of the tower, and Alan was forced to ship more malleable pink granite from Mull. Even the ship itself caused trouble. The Commissioners originally balked at buying a specially built tender solely for use on the Skerryvore works, but finally relented after it became evident that no other ship could withstand the Hebridean seas and Alan's demands. Until the Skerryvore tender was built, the workmen endured the seasick lumberings of the main lighthouse ship, the *Pharos*. 'The inconvenience arising from her heavy pitching was, to landsmen, by no means the least evil to be endured,' Alan noted later.

Planning the works was enough to keep him occupied for months, and the lighthouse itself was almost subsumed under paperwork and arrangements. But the design of the light had occupied long days of thought and worry. In theory, he could simply have taken his father's design for the Bell Rock and adapted it to Skerryvore's conditions. In practice, he began again from scratch. Skerryvore, with its fissured rock, its exposure to the force of the Atlantic, and its series of creeks and snags, made the Bell Rock seem easy. Waves pounding the reef could exert up to 6,083 lbs of liquid pressure, a force of nearly three tons per square foot. On the Bell Rock, by comparison, the highest measurement had been 3,013 lbs of pressure, around one and a half tons. If the equation was extended to the full height of the tower, it had to be capable of withstanding a force of several thousand tons. It seemed an exaggerated calculation, but, as the lightkeepers later confirmed, unbroken water did frequently soar over the lantern during heavy winter storms, shattering the panes and making the tower shake beneath their feet.

Alan, therefore, had to find the best method of constructing a tower whose shape and weight resisted the sea in the most efficient way possible. His first step was to ensure that the centre of gravity was as low as possible; his second was to distribute the weight of the stones to optimum effect, and his third was to build high enough to give the waves least chance of leverage. With all of this in mind, he designed a tower 138 feet high and weighing 58,580 tons, with nothing that the sea could get a grip on but a narrow cornice at the base of the lantern. The Bell Rock, by comparison, was 100 feet high, and 28,530 tons in weight; the Eddystone 68 feet and 13,343 tons. His final choice was an aesthetic one. Having decided on the most practicable design for a light, he was left with four possible formulas for the shape of the tower. He chose a hyperbolic curve because it pleased his eye. It was, he said, 'a simple and almost severe style'. He designed it perfectly, but also made it beautiful, just for the sake of it.

In one sense at least, Alan's plottings were not conducted alone. All his choices were made with two influences in mind: the ghost of Smeaton and Robert. Alan proved himself strong enough for both of them, disputing Smeaton's wisdom that any sea-tower should be built on the principle of an oak tree, and dispensing, through shrewd physics, with his father's dovetail joints. Alan reasoned that the sheer weight of the tower coupled with the even distribution of its tonnage would dispense with the need for dovetails in all but the upper courses. In Skerryvore's case, none of the tower would be underwater at high tide, so there was little fear of water dislodging the masonry while it was being constructed, and Alan, conscious of making savings where possible, concluded that the jigsawed joints at the Bell Rock were both expensive and unnecessary on Skerryvore. Considering Robert's involvement in the planning of the light, it must have taken some courage for Alan to defy his father so directly. Robert kept a keen eye on Alan's progress throughout the works. Alan pacified him with honesty, diplomacy and the occasional sin of omission. He did, however, draw upon his father's work in other respects. Robert's idea of building a temporary workmen's barracks on the rock to keep stores and give shelter to the workmen seemed to Alan to be essential on Skerryvore where the foul seas often delayed the boat trips back to Hynish. 'So important, indeed, did this object appear to me, that I was at times apt to look upon it as an indispensable step towards ultimate success,' Alan confessed. The barracks was to be built on much the same lines as his father's version – six sturdy wooden pillars trussed together with iron bands supporting a rocket-shaped cabin.

On 5 May 1838 Alan sailed out from Greenock with a cargo of materials on his way to Hynish and the start of the works. Some progress had already been made. Sixty fine blocks of Tiree gneiss had been quarried, the workyard was almost completed and a gunpowder magazine had been prepared. Alan spent a month at Hynish checking stores, finding new sources of stone and instructing the workmen. It wasn't until 28 June that Alan finally reached the rock, and was left there while the row-boat returned to the tender for the rest of the working party. Stuck alone on Skerryvore and watching the waves lap higher with the tide, Alan felt a thrill of morbid fear. 'I could not,' he wrote later, 'avoid indulging in those unaccountable fancies which lead men to speculate with something like pleasure upon the horrors of their seemingly impending fate. These reflections were rendered more impressive by the thought that many human beings must have perished amongst these rocks.' Alan was not usually inclined to superstition, but no one who saw Skerryvore could have remained entirely dull to its history and reputation.

When it did finally begin, the season's work progressed in lurches and bursts. One day's good weather would be succeeded by five days' bad. Alan calculated that between 7 August and 11 September only 165 hours of preparatory work had been completed. The rest of the time was spent storm-stayed in Hynish, sailing from Skerryvore to Tiree or waiting on the rock for stores to arrive. Everything was conducted in an atmosphere of terrible flustered tedium; endless time spent staring at gales interspersed by periods of frantic activity. During fine days, the men worked from 4.00 a.m. with a pause for breakfast at 8.00, then from 8.30 until 2.00 p.m., with another pause for dinner (supplemented by vegetable broth and a little beer), then worked on until 8 or 9.00 p.m., when they would scramble for the boats and the journey home to Hynish. Supper on the boat was rarely eaten with much relish, since the workmen were exhausted and suffered from constant seasickness. The addition of a salted sheep to their provisions did not much encourage their appetites.

Alan sympathised with the men's complaints, but took consolation from their enthusiasm for argument. 'The amount of hard labour and long exposure to the sun, which could hardly be reckoned at less than 16 hours a day, prevented much conversation over supper: yet, in many, the love of controversy is so deeply rooted, that I have often, from my small cabin, overheard the political topics of the day, with regard to Church and State, very gravely discussed on deck, over a pipe of tobacco,' he noted affectionately. When it was necessary to stay overnight on the tender, the men would spend a sleepless night pitching in their bunks before rising at 3.00 a.m. and landing on the reef. Often, the boat would be unable to get near Hynish, and would be forced to sail aimlessly round the coasts of Mull and Coll, looking for a safe mooring. Many preferred to eat all their daytime meals on the rock rather than risk another bout of nausea on the boat.

Once on the rock, the work progressed speedily. When a few of the supplies had been landed – including a smith's forge, two anvils, quarrying equipment and shear poles for raising heavy beams – Alan plotted out a place where the barracks could stand. Boreholes were chiselled into the gneiss for the foundations of the barracks and the iron legs shaved and slotted into the rock. Eventually, after a great deal of heaving and pulling, the six giant poles were levered upwards and bound firmly into place. Most of the operation had been spent battering away at the rock, which was so striped with seams and flaws that it was almost impossible to make any straight cut. Alan had also made a small wooden hut to shelter the forge and landed as many of the materials as he felt could safely be stored on the rock. On 11 September, having battened down the barracks as securely as possible, he and the workmen left the rock for the winter.

He returned briefly to Hynish, congratulated everyone on their achievements and sailed southwards 'in the pleasing belief that the successful termination of our first season's labours might be taken as an omen of future success'.

Two months later, while back in Edinburgh, Alan received a letter from Mr Hogben, the storekeeper at Hynish. 'Dear Sir,' it began, 'I am extremely sorry to inform you, that the barrack erected on Skerryvore Rock has totally disappeared.' It had been visible on 31 October, but the three following days had brought a driving west coast rain during which no one had been able to see the rock from Hynish. On the evening of 3 November, 'the wind increased to a gale, with a great swell, and an extra-ordinary high tide. Yesterday (Sunday the 4th) ... Mr Scott and Charles Barclay ... got a momentary glimpse of the Rock through the spray, and both were of opinion that the barrack was gone. This was not credited by the workmen who had been employed at it, but this morning we found it to be the case; the Rock was pretty clearly seen, but no trace of the barrack.' A whole year's work was lost; a season's labour flung to the winds in one day's gale. Alan took sparse comfort from Mr Hogben's assurance that the islanders hadn't seen such a severe storm for over sixteen years. As he later pointed out, 'Thus did one night obliterate the traces of a season's toil, and blast the hopes which the workmen fondly cherished of a stable dwelling on the rock, and of refuge from the miseries of sea-sickness, which the experience of the season had taught many to dread more than death itself.'

That same evening, he set sail for Tiree. The weather was still too foul to land on Skerryvore, but from the boat's deck he could see the destruction for himself. It was, he found, just as Mr Hogben had specified. The beams supporting the barrack had disappeared, the iron stancheons had been ripped from the rock or twisted beyond repair, the smith's forge had vanished, and the anvil had been thrown about eight yards to the south-west. The mason's tools, the quarrying equipment and several moulds had either been smashed to pieces or stolen by the sea, and the grindstone had been hurled from the top of the rock into a gully at the bottom. The two crabs (small cranes, or derricks, for lifting) had both been dashed on the rocks, the mooring buoy had gone and a large iron ring used to haul weights had been yanked from the rock into the sea. A loose boulder measuring three-quarters of a ton had been shifted from the base of the rock to the top. Alan stared at the fragments, and then turned away. The journey back to Hynish, 'in all the gloom of a stormy night, and depressed by mingled disappointment and sad forebodings', was not an easy one. The workmen, he noted, were 'very sick and much dispirited', and seemed almost as upset by the disaster as Alan himself. Alan returned to Edinburgh, and wearily began redesigning his plans.

The construction of the remodelled barracks, strengthened, bolted and anchored to a different site on the rock with every last beam and nail that Alan could find, occupied a good part of 1839. Almost nothing of the previous season's work could be salvaged except for the bore-holes made in the rock and a few pieces of flotsam that floated ashore at Hynish some time later. All that remained of the old barracks, as Alan noted, was part of a beam attached to one of the iron stancheons, 'so thoroughly riven and shaken as to be quite like a bundle of lathwood'. Meanwhile, Alan busied himself with cutting and moulding the stones from the new quarry on Mull, blasting the foundation pit for the light and constructing a proper mooring on the rock. These tasks also took almost a whole year.

The new pink Mull granite might have been more malleable than the Tiree gneiss, but the business of blasting, dressing and finishing the stones, and then shipping them twenty-six miles to Hynish was still a painstaking job. At the quarry, the workmen bored holes in the rock, primed them with powder like an old musket gun and then bolted for the nearest refuge as several hundred tons of granite exploded behind them. Once prepared, each block weighed between three-quarters of a ton and two and a half tons, and even when Alan had dispensed with his father's dovetailed cuts they still required careful cutting and shaping. He was perfectionist enough to insist that no stone should be more than an eighth of an inch out of line. As he pointed out to Robert, he wanted to complete 'every work of a preliminary nature in as substantial a manner as possible so as to secure to ourselves the advantage of entering next season unencumbered by the drawbacks of other works upon the two great operations of landing and building.' Even so, such pedantic standards meant slow progress. Alan later calculated that it had taken around 120 hours to dress a single stone for the outside of the tower, and 320 hours to dress a single one of the central stones. In total, 5,000 tons of stone were quarried, shaped and shipped. The whole operation lasted almost a year and a half.

Once out on Skerryvore, those men spared from quarrying or barracks-building were preparing the foundation pit. 'A more unpromising prospect of success in any work than that which presented itself at the commencement of our labours, I can scarcely conceive,' Alan wrote, gloomily. 'The great irregularity of the surface, and the extraordinary hardness and unworkable nature of the material, together with the want of room on the Rock, greatly added to the other difficulties and delays, which could not fail, even under the most favourable circumstances, to attend the excavation of a foundation-pit on a rock at the distance of 12 miles from the land.' As he pointed out, the stones of Skerryvore were four times as hard as that old lodestone of resistance, Aberdeen granite. Producing a levelled pit forty-two feet in diameter from a reef with fewer flat surfaces than the Cairngorms was not something any engineer, of whatever skill or brilliance, would relish.

After much deliberation, Alan selected the only site that seemed feasible and instructed the quarriers to begin digging. The process was much the same as for cutting the stones: drilling deep holes, priming them with powder, and retreating to a safe distance. Except that on Skerryvore, there was no safe distance; there were only a few yards of rock to either side, and then salt water as far as the eye could see. Alan took what precautions he could, including covering the mine-holes with netting and giving all the workmen protective fencing-masks. But it was not a comfortable time, being stuck on a hostile rock, surrounded by whirling shrapnel and a group of terrified workmen dressed like sword-fighters. Most alarming of all was the Atlantic itself. Gales on the east coast were at least prefaced with glowering clouds and bucking seas; out on the west, there would be a sudden, ominous lull, a spatter of rain, and then a wind strong enough to lift a grown man bodily off the rock. Alan and the workmen often found themselves bolting for the boats, leaving tools, materials and provisions scattered where they lay. Small wonder that when they returned to the dubious safety of the boat and the long journey home every evening, most of the men could be heard mumbling prayers. 'Isolation from the world, in a situation of common danger, produces amongst most men a freer interchange of the feelings of dependence on the Almighty than is common in the more chilly intercourse of ordinary life,' Alan noted wrily.

Once the surface of the rock had been levelled, the masons began picking out a pit. The foundations did not have to be deep – no more than fifteen inches – but it was still finicky work, particularly when the gneiss could blunt a pick in three blows. Alan wanted the lip of the pit to fit so closely to the stones of the tower that nothing – no wave, no stones, no insinuating rain – could slide between the two. After 217 days, the foundations were finally completed. Several of the workmen were so proud of what they had achieved they declared themselves sorry to see it covered up. Alan confessed modestly to his father that 'I have carefully examined the whole of the foundation Pit which is now complete, and a beautiful piece of work, I must pronounce it, though perhaps I ought not to say so . . . I believe there is no such foundation in Europe for hardness and fineness of surface.' In fact, it pleased him so much that he threatened to spend an extra night on the rock, 'not because my presence there is particularly necessary for any special purpose, but simply because I have seldom visited it with so much satisfaction.' Fortunately, the Hynish doctor dissuaded him.

The final duty of the season was to fix and carve out a mooring place. Every morning the workmen still made the hazardous scramble from the row-boat to the rock, carrying tools, materials and ungainly cranes up the sides as best they could. One man bitterly compared the slippery climb up the reef to shinning up the neck of a bottle. Since the best place for a landing site was underwater at high tide, it needed to be blasted quickly and decisively. Alan's main pleasure was in watching the islanders' reaction to the dynamitings.

> After all the mines were bored and charged and the tide had risen, and every one had retired from the spot, the whole were fired at the same instant, by means of the galvanic battery, to the great amazement and even terror of some of the native boatmen, who were obviously much puzzled to trace the mysterious links which connected the drawing of a string at the distance of about 11 yards, with a low murmur, like distant thunder, and a sudden commotion of the waters in the landing-place, which boiled up, and then belched forth a dense cloud of smoke; nor was their surprise lessened when they saw that it had been followed by a large rent in the rock, for so effectually had the simultaneous firing of the mine done its work, that a flat face for a quay had been cleared in a moment.

A gale followed soon afterwards, stripping the rock of several timbers, the smith's bellows and all the moorings, including a ladder and several wooden fenders. It was the second set of moorings that the sea had stolen. Alan took it as a further sour omen that Skerryvore did not mean to give in easily.

On 3 September, Alan left the reef and returned to Edinburgh. Despite his preoccupation with the works, he had still maintained a regular correspondence with his father, and was expected to give a full account of all his operations to the Commissioners. Robert and the Board expected Alan to justify everything from the cost of chisels to the workmen's diet. Every penny was counted up, counted round and counted back over and over again. He also managed to arbitrate in a railway dispute, draw up plans for riverworks in Preston and contribute a sizeable entry on sea lights to the *Encyclopaedia Britannica*. In his absence, David had moved shrewdly into position within the family business, finding the office 'greatly in want of someone to devote to it his *undivided and constant* attention'. Robert had made him a full partner, and entrusted him with much of the work for the Convention of Scottish Burghs. Tom, meanwhile, was in Cardiff, learning how to be an obedient engineer. The winter, though a respite from the working season, was no rest for Alan. The long nights moored on the tender or in the workyard at Hynish had at least offered a brief solitude; Baxter's Place, under his father's fretful attention, did not.

Work on Skerryvore began again in April 1841, and for the first time the men were able to abandon the lighthouse tender for the barracks. It had lasted the seven months of winter largely unscathed, though Alan noticed that five tons of rock had been cleared by the sea from the foundation pit. Once stores were landed, the men moved in. They rapidly discovered that they suffered just as much in the barracks as on the boat. It had been fitted out with hammocks, a small cooking stove, fresh-water tanks and a few provisions, but the space was so cramped that there was nowhere to store anything but essentials. When it rained the men stayed wet as the barracks was too small to store spare dry clothes, and during heavy rains, the rooms were often flooded with water. When a storm started, Alan and the men endured a comfortless few days trapped in the barracks, waiting and watching. Since the rooms couldn't be heated, they spent most of their time in bed,

> listening to the howling of the winds and the beating of the waves, which occasionally made the house tremble in a startling manner. Such a scene, with the ruins of the former barrack not 20 yards from us, was calculated only to inspire the most desponding anticipations; and I well remember the undefined sense of dread that flashed across my mind, on being awakened one night by a heavy sea which struck the Barrack, and made my cot or hammock swing inwards from the wall, and was immediately followed by a cry of terror from the men in the apartment above me, most of whom, startled by the sound and tremor, immediately sprang from their berths to the floor, impressed with the idea that the whole fabric had been washed into the sea.

At one point, the weather was so bad that they were stuck on the rock for two full weeks. By the time the sea calmed down enough to allow the supply boat to reach them, they had only a day's worth of food left.

Alan spent his evenings confined to a tiny cabin in the barracks, writing letters, keeping up his journal and replenishing his need for solitude. He kept a small store of books and confessed later that during long nights and difficult gales he had time to read *Don Quixote*, Aristophanes and Dante twice through. By now, he had struck up a correspondence with William Wordsworth, and was later to visit him in the Lake District. The friendship had been established through James Wilson, a zoologist in Edinburgh. In 1841, Wilson, a long-standing friend of Wordsworth's, had sailed round the coast of Scotland on a tour of inspection with the Board of Fisheries and had visited Alan at Hynish. In December 1842, Wilson wrote to Wordsworth introducing Alan and Skerryvore as one of Scotland's more intriguing native curiosities. According to Wilson's assessment, Alan was 'an honest man, of good natural intelligence, but with no pretension to literature'. Alan, he said, had written to Wilson describing his delight at receiving several mementoes from Wordsworth – a lock of his hair, a few laurel leaves from the poet's home at Rydal Mount and his autograph – and explaining the pleasure and sense of endorsement he gained from Wordsworth's poetry. 'I am highly gratified to find that the Poet should receive any satisfaction from the testimony of a remote and obscure *doer* of daily work like myself to the value of his writings,' Alan wrote. 'Many were the moments in my solitude, during which I have felt my commonplace labours ennobled by the Poet's views of duty and perseverance . . . while other poets unfit us for an immediate relish of every day labour by carrying us into situations wholly chimerical, Wordsworth so elevates the mind by presenting common objects in their true relations to Man's more exalted hopes and fears, as to dignify all his worthy thoughts or occupations however humble.' He had, he confessed, stuck the poet's mementoes alongside small portraits of Southey and Coleridge on the wall of the barracks to sustain him during the days of work and worry. Alan later kept up the correspondence with Wordsworth, unassisted by the well-intentioned patronisings of Wilson, and the two found common ground in their discussions of poetry, criticism and nature.

Many years later, Wordsworth's son Gordon wrote to Alan's daughter Katharine de Mattos correcting Wilson's estimate of Alan. 'Prof Wilson may be right in saying that your Father had no literary *pretension*,' he pointed out, 'but it is certain he had very remarkable literary *insight* and power of expression. It is notoriously difficult to appraise contemporary poetry rightly – even Coleridge was often wrong. Your father seems to me to have been the first to give the same verdict as the most distinguished admirers have since done in no better words, critics like Mill, Matthew Arnold and Lord Morley.' As Wordsworth himself had noted, acknowledgement from Alan, ostensibly a non-literary figure, was 'more valued by me, far more than others I receive, both as testimonies to the substantial benefit drawn from my writing by Contemporaries and as presumptive evidence that the same benign operation will go down with them to succeeding times.'

The letters from Wordsworth and others were some of the few outside intrusions in an otherwise hermitic life at Skerryvore. When the weather was good, all the men laboured like navvies. Alan noted proudly that not more than half an hour of working time had been lost during the whole of their time on the rock. 'I am inclined to believe that throughout this summer in particular, the men have been working close upon the limits at which human strength begins to fail, and I have myself repeatedly fallen asleep in the forenoon with my pen in my hand,' he reported to the Commissioners. When he was awake, much of his time was spent watching the sea and the sky. For a while, he tried to measure the waves but couldn't find a way of fixing a measuring pole to the sea bed. From the desultory observations he could make, he discovered that the waves rarely reached higher than fifteen feet, though the sight of a wall of blue-black water towering above them usually sent even the most seasoned workmen hurrying for shelter. Alan found the conditions both exhausting and exhilarating. Sometimes, distracted from the hammering of the masons, he would catch sight of bright shoals of strange fishes, or find a freshly laid gull's egg nestled between crevices in the rock. He was not interested in ceremony, as his father had been, but took satisfaction from more elusive things: the stark beauty of the sea and the pleasures of rest after hard physical labour.

Considering the dangers the men were expected to deal with, there were remarkably few accidents at the works. George Middlemiss, foreman of the carpenters, died halfway through the project, his heart inexplicably paralysed. One workman died of consumption. One of the masons became violently ill during his first night on the barracks, and his screaming spasms throughout the night prevented his colleagues from sleeping; all of them were pleased to see him rescued and borne off to Hynish the next day. Alan had employed a full-time doctor for the 150 workmen, but was lucky to escape with a thin tally of seriously injured men.

With the working conditions as bad as they were, however, it was unsurprising that some of the workmen objected. The previous year, the boat crew had mutinied, complaining that Alan's insistence that they moor so close to the reef was both dangerous and absurd. They insisted that if they did not get an immediate increase in wages, they would abandon Alan and everyone with him on the rock. The crew's argument did not impress the steamer's master, who dismissed them as soon as they were able to return to Hynish. But the mutineers' sour temper affected the rest of the workmen and there were several desertions throughout the course of the works.

There were other difficulties as well: complaints over wages, conditions or the liquor ration, and arguments between the men that grew overheated. The case of Mr Hill, one of the foreman masons, was unusually troublesome. Hill's behaviour had been 'the talk and merriment of the country', Alan reported to Robert. He had 'been guilty of excessive folly in regard to giving himself airs, being quarrelsome and occasionally intoxicant. The Doctor gave me a woeful account of his absurdity and is of opinion that he is *touched*. I made every enquiry I could and find it too true. I therefore gave Mr Hill notice to prepare to leave from all together along with me. He was at first violent and high; but I told him calmly but resolutely that my mind was made up and that he had better submit calmly than make any noise . . . You need not,' he added in a hurried postscript, 'show this to my mother as it may make her uneasy. She has a great fear of, "daft men"!' Small wonder poor Jean Stevenson worried. Her son was stuck alone on a lonely rock, confined to a tiny barracks and forced to live under military discipline with a group of potentially mutinous workmen. In such a situation, any mother would have worried for her son. Things were particularly dangerous when the workmen consoled themselves with smuggled whisky, as they often did, and ended up flailing at their colleagues.

Once the foundation pit had been levelled, Alan began landing the stones for the first few courses. He found, just as his father had, that trying to land a two-stone lump of granite from a rearing boat onto a slippery reef was an uncomfortable task. The crew had to get as close to Skerryvore as they dared, chain the stones, hook them to the crane and wait warily while the swinging rock was hauled up. If they came too close to the rock, they risked being lifted by a wave and thumped down on the reef; if the crane lost its grip, the men would have been flattened by the stone's fall. The sea was often so bad that the crew had to be lashed to the boat's railings, and on one occasion, a sudden swell snapped eight mooring ropes at once, knocking the crew flat and shoving the boat violently to one side. Eight hundred tons of rock were finally landed that season and every piece had given Alan and his crew another jolt of fear.

On 7 July, the Duke of Argyll and his entourage came to lay the foundation stone and inspect the progress of the works. Alan was flattered by their interest in the works, but irritated that their visit would mean losing a good hour's work through needless ceremonials. Afterwards, he wrote disconsolately, 'Never did any day prove more unfortunate than yesterday . . .after the Duke, Duchess, his brother-in-law, Mr Sinclair, the Marquis of Lorn and his sister Lady Emma Campbell were all safely shipped and we had gone out a few miles we were forced to turn back at the Duchess' instigation as she became alarmed. I was truly sorry for the poor Duke who resisted returning as long as he could and yielded most unwillingly as he was quite bent upon the whole enterprise.' After the Duchess had recovered from her nerves, the Duke returned, this time only accompanied by his factor, and they 'laid the first stone out with no ceremony but three cheers and a glass'.

The first four courses of black Tiree gneiss were laid with relative speed. Alan's insistence on getting the blocks cut exactly to size paid off; within fourteen hours it was possible to lay eighty-five stones, and by the end of the season the tower had risen eight feet from its foundations. Much of the skill was in the preparation: each custom-cut stone was carefully allocated its own slot, none could be substituted for another, and every course fitted into its neighbour as neatly as the rings of a tree. As the work progressed, Alan found himself increasingly aware of Smeaton's influence, and confessed to his father that 'I now feel doubly anxious to visit the Eddystone, having laid my first course on a foundation which resembles Smeaton's in many respects.' He was aware that he could not afford to show weakness even now; everything Smeaton and Robert had done in their time he had to match or better. But it was more than just his father's voice goading him, it was his own gathering self-belief, won through time and strength. At one stage, trapped on the 'desert rock' for five weeks, Alan discovered to his astonishment that he was actually enjoying himself.

The whole of the 1841 and 1842 seasons was spent landing, checking and fitting the stones for the light. It was heartwarming work. Most of the preparations for building had been so thorough that the tower was laid and fixed with spectacular rapidity. Some time was spent fitting up cranes to lift the higher stones to the top of the tower, the rest was spent mixing lime and pozzolana earth to give the strongest mortar possible. The fit of the stones was so perfect that Alan subsequently discovered only two joints where leaks had developed. Even more satisfactory was the discovery that each course only diverged by a sixteenth of an inch from his original calculations. He wrote happily to Robert in July 1841, 'It gives me great pleasure to say that one more lighter load will land all the Solid on the Rock and that we shall then have 27,110 cubic feet of masonry on the Skerryvore, a quantity little short of the whole mass of the Bell Rock and more than double that of the Eddystone. We have been going on with great spirit. Double trips daily, or about 1,000 cubic feet a day landed on the Rock.' Not that any of them could afford to be complacent; when work began again in 1842, Alan noticed that stones had been hurled from the sea into the unprotected top of the tower, over sixty feet above high water. In bad weather, the sea would snatch at the stones laid out at the base of the tower, and the workmen often had to mount midnight rescue operations to lift the stones out of the water's reach.

On 21 July 1842, the last of the stones were landed. Four days later, the tower was finished. All that remained was to fit the great iron lantern, hoist up the lenses and furnish the rooms. The lens was another of Alan's innovations, negotiated in great detail with Augustin Fresnel and his lens-maker, Soleil. It was to be a revolving dioptric lens apparatus, with large inclined mirrors spread fan-like above a hexagonal lantern. Circular lenses in front of the argand lamp focused the rays, while mirrors and pyramidal lenses caught any remaining light and drove it outwards, producing a beam so strong that Alan proudly noted that it could sometimes be seen from the heights of Barra, thirty-eight miles distant. The specifications were complex enough to prevent Soleil shipping the lamp for a further two years. It was Tom, in the end, who completed the Hynish dock and fitted the lens.

But the tower itself was finished. Once fitted out, it lost a little of its rigorous simplicity but none of its elegance. A gun-metal ladder curved from the base to the door. Above the door, there were a further eleven floors, including that of the lightroom, two small subdivided floors (sleeping quarters for the keepers), a kitchen fitted with an immense black cooking range, a room for visiting officials, a workroom and a series of provision stores. Visitors climbed a tiny wooden ladder through a hatch between the floors, and the keepers hoisted stores up and down on their backs. There was no bathroom: keepers had to make do with the freshwater tanks, an iron bucket and a prudent diet. Alan suggested to the Commissioners that all future keepers at Skerryvore should be paid a £10 premium on their salaries, to compensate for the 'peculiar disadvantages' of the place. The Bell Rock keepers, by comparison, lived a life of luxury and high society. At Skerryvore, 'the situation of the lightkeepers will be one of almost complete banishment even from their families during several months in winter.' This was no idle claim. During 1843, while a party of workmen were fitting up the lantern, the weather remained so bad that the boat could not leave Hynish for seven weeks. 'The poor seamen who were living in the Barrack passed that time most drearily,' Alan wrote later, 'for not only had their clothes been literally worn to rags, but they suffered the want of many things dearer to them than clothes, and amongst others of tobacco, the failure of the supply of which they had despondingly recorded in chalk on the walls of their prison house, with the date of their occurrence!'

The lighthouse on Skerryvore was, by any standards, an astonishing piece of engineering. It consisted of 137 feet of granite, weighing a total of 58,580 tons, with walls at the base nine and a half feet thick. When finally completed, it had taken seven years, £90,268, 150 workmen and the best part of Alan's working life. It was, as the Institute of Civil Engineers put it, 'the finest combination of mass with elegance to be met with in architectural or engineering structures'. Louis summed it up a little more pithily as 'the noblest of all extant deep-sea lights'. It still is. Its making had been a form of self-imposed exile, absorbing every energy of body and spirit for all of those involved. In the time that it had taken to build Skerryvore, Queen Victoria had succeeded to the throne, Hong Kong had been appropriated by the British, the Corn Laws were fought and the Great Famine in Ireland had begun its long slaughter. At home, Tom had grown up and Robert grown old. Alan had built his masterpiece and had almost destroyed himself in the process. He went home, back to life and love and other lighthouses. After Skerryvore, it was not the same.

SEVEN

Muckle Flugga

The home that Alan returned to had changed in his absence. During the tense years of Skerryvore's construction, his brothers had both risen to fill the gap he had left. David now assumed almost complete control of the family firm and Tom, having finally qualified as an engineer, was assuming more responsibility for the lights. Robert's assistants, now long past the days of their apprenticeships, were taking on projects of their own or were involved in supervising the Commissioners' business. Baxter's Place was now the hub of an organisation with interests in almost all of Scotland's engineering schemes, a place where the shape of the nation was traced out afresh every day. With Robert's sons all fully grown, the lights and the business were no longer the territory of one man. The days of the pioneers had gone, and engineering was becoming more a matter of business than adventure.

As his sons grew, Robert faded. On 16 November 1842, he wrote a painful letter to the Commissioners announcing his decision to retire. He was, as he conceded, now seventy, and no longer capable of fulfilling the role of chief engineer. 'It is not without regret,' he said, 'that I now find that my advancing years and the daily extending and important duties of my Office warn me of the necessity of retiring from public duty ... In 1797 I made my first survey of the entire Coast by direction of the Board and I have ever since Continued to perform this often perilous duty and to make the annual Reports and Requisitions for the Service of the Year.' As he pointed out, his record with the Northern Lights was exceptional by any standards. During his tenure he had been responsible for the expenditure of almost £1 million, written over three thousand letters a year and been at work for the Commissioners for almost half a century. 'I have only to express a hope that the Board will duly bear in mind the long continued and I may be pardoned for adding zealous attention with which I have served them up to the present hour; and that when they come to consider the Subject of my retiring allowance they will not forget the very peculiar nature of my services, particularly its responsibility during the early part of my Career as their Engineer, and further that I have brought the Reflecting Lighthouses from their former imperfect state to their present perfect Condition.' In reply, Sheriff Maconochie noted 'at some length' their appreciation for Robert's services, regretfully accepted his resignation, and acknowledged that 'his skill, attention and zealous anxiety to promote the welfare of the Establishment is in a great measure to be attributed the present admirable system in which the Lighthouses under their Care are now Carried on.'

Robert was not so diminished that he gave up work entirely. He had been involved with the lights for so long that he found it impossible to imagine life without them. For half a century, the pattern of his life had been dictated by the annual inspection voyages and the plans for new projects. The lighthouses were his job, but they had also become part of his soul. Besides, it was entirely beyond his nature for Robert to resign himself to a slow retirement. In the minds and Minutes of the Commissioners, he might have appeared as a public functionary, but to the lightkeepers, masons and sailors, he *was* the Northern Lights. Every inch of Scotland's coastline was stamped with his knowledge and character; every light in Scotland had a small part of Robert Stevenson built within it. He and his family were indivisible from their work, and no amount of official procedure could separate the two.

The years of Robert's retirement were therefore accompanied

by a further flurry of work. He continued receiving petitioners, vetting new assistants and guiding the direction of his sons' work. He despatched Tom off on a survey of all British harbours, presenting him with a densely written notebook filled with tips and hints. He started dictating a history of his life to his daughter, Jane. He began a scrapbook filled with cuttings of useful and illuminating articles (notices of railway accidents, river improvements, accounts of storms and statistics for drunkenness in Edinburgh), and often grew so impatient of journalistic inaccuracies that he took to editing the clippings himself. He stepped up his involvement in the Royal Scottish Society of Arts (an institution devoted to the dissemination of 'useful knowledge amongst the industrious classes') and wrote a couple of papers for the learned journals. The lights, as always, continued to preoccupy him. In 1844, he visited Skerryvore twice, ostensibly in his capacity as private 'civil engineer', but both times travelling on the lighthouse tender. David, Tom and Alan had all become accustomed to his habit of overseeing their work. After his retirement, the number of notes, instructions and peremptory letters they received from him only increased.

Robert also began a troubled correspondence with the Church. This was the age of the Disruption, the decisive split between the Established Church of Scotland – governed by a conservative mixture of civil and political forces – and the Free Church of Scotland, which had broken away in order to be free from state interference and to choose its own ministers. Robert, a staunch believer in the Established Church, found the split upsetting. When his local minister wrote suggesting that he might wish to come to a meeting of the kirk session to decide whether to go over to the Free Church, Robert wrote back brusquely declining. 'As I do not at all go along with these proceedings, it does not appear to me that my attendance would be for edification.' Age had only made him more conservative, and he was repelled by the threat of upheaval, whether in religion or in his private life.

Alan, meanwhile, had no difficulty at all in filling his time after Skerryvore. In the same meeting that Robert's letter of resignation had been read before the Commissioners, he submitted his application for the post of chief engineer. The professional record he laid before them was impressive enough; including two separate degrees (as Master of Arts and Bachelor of Laws), the apprenticeship with Robert and then Telford, the construction of eight major lights, the introduction of dioptric lenses within the Scottish lights, the 'numerous laborious and intricate computations', for light sources, the private work on rivers, harbours, roads and bridges, and, most conclusive of all, the design and execution of Skerryvore. After some token deliberation, the Commissioners appointed him successor to Robert on a salary of £900 per annum. The only flaw in Alan's promotion was the workload it entailed. The Commissioners saw no urgency in appointing another clerk of works, so Alan was left to undertake two jobs. In addition, he was expected to write a full account of the Skerryvore works for general publication, just as his father had done for the Bell Rock two decades before. Admittedly, Skerryvore had not attracted the same instant public acclaim that the Bell Rock had done. Alan was a quieter man than his father and lacked the added boost of Sir Walter Scott's patronage, but his fellow engineers considered that his achievements had outstripped Robert's work. An admiring notice in the *Quarterly Review* declared that the great triumvirate of lights – the Eddystone, the Bell Rock and Skerryvore – 'are perhaps the most perfect specimens of modern architecture which exist. Tall and graceful as the minaret of an Eastern mosque, they possess far more solidity and beauty of construction; and, in addition to this, their form is as appropriate to the purposes for which it was designed as anything ever done by the Greeks, and consequently meets the requirements of good architecture quite as much as a column of the Parthenon.'

Alan was much more at home with writing than his father, who had found the preparation of his book an elaborate torture.

But still, a full account of the works required time and patience to prepare. Alan was pushed into writing the book in spare moments at sea or during inspection tours, as he admitted in the preface, 'My labours were also continually interrupted by the urgent calls of my official duties; and, on several occasions, I was forced to dismiss unfinished chapters from my mind for a period of several months.' He, like Robert, was continually being despatched to London to answer to parliamentary committees or to the Trinity House men. Unlike Robert, he was not a political animal, and undertook the public side of his lighthouse work dutifully but without relish. Robert's talent had been to keep all of his interests alive at once, rushing from private business to political meetings with his usual energy. Alan's talent was to focus on one object – the refinement of lenses, the writing of papers, the building of lights – with absolute intensity and thus to make it great. He was uncomfortable with self-advertisement and generally preferred his work to speak for itself. As the Commissioners noted approvingly, his style might have been very different from his father's, but what he brought to the lights was of equal value.

The Commissioners also expected him to have all Robert's energy, and more. Between 1844 and 1854, he was responsible for the design and construction of ten new lights, double what Robert had been asked to achieve in two decades. Since he was still also acting as clerk of works, he was responsible for ensuring that every detail of those lights, from the shape of the cornicing to the slates in the roof, was checked and executed correctly. The NLB was now in charge of over thirty lights, but there was still no set formula. Every light was custom-designed to fit the local conditions. At Ardnamurchan, for instance, Alan was able to use the same Mull granite that he had used on Skerryvore, but, since it was a light built on a headland, the design of the tower was entirely different from the Skerryvore model. Ardnamurchan is a hummocky, rock-scarred peninsula reaching out to the westernmost tip of the British mainland and overlooking the dangerous waters of the Minch. There is an old mariner's tradition that any sailor who gets past the Point of Ardnamurchan safely is entitled to place a sprig of white heather on the mast of his boat as evidence of his good fortune. For this spot, Alan drew up plans for the only 'Egyptian style' lighthouse in Britain, with a graceful arched cornice and gently tapered walls. Round the base of the catadioptric lamp, he sketched in stylised Egyptian figurines. One of Alan's most endearing characteristics was his love of godly details – a discreet flourish on a cornice, a brass lion's head on a lamp stand, sea-serpents on a lantern or claw-feet for a base. The one tiny bloom of artfulness on an otherwise functional lighthouse or lantern was both Alan's signature and his wistful hint of another life unlived.

The tests of the new dioptric lenses – first at Inchkeith and then, with improvements, at Skerryvore – had proved so successful that Alan's next project was to extend them to as many of the lights as possible. By 1847, lenses had been installed in seven of the fixed lights, and almost half of the revolving lights, and, aside from occasional problems in persuading suspicious keepers to accept the new machinery, all were working well. Most gave a light three times as powerful as the old reflectors, capable of being seen from up to thirty miles away.

His work with the lenses obsessed him almost as much as Skerryvore had done. In a revealing aside, he claimed that 'Nothing can be more beautiful than an entire apparatus for a fixed light of the first order . . . I know of no work of art more beautiful or creditable to the boldness, ardour, intelligence, and zeal of the artist.' At one stage, he became so preoccupied with the performance of his lenses that, when a ship was wrecked on the Isle of May, he interviewed the surviving crew in order to discover whether the light itself had somehow been at fault. Both the ship's master and mate admitted that the light had been perfectly clear, but they had become trapped on a lee shore. Alan noted that their experience was a 'most satisfactory conclusion'. Public approval was more ambivalent. Some complained that the light was now too bright; others that the expense (around £450 for each new set of lenses) was absurdly high. Doomsayers, however, had always existed. When Robert had changed the light at Inchkeith from a fixed to a revolving beam, one elderly woman who could see the light from her home in Leith remarked that she felt sorry for the keeper. She saw the beam flashing, and believed that he was forced to sit in the lantern all night relighting it every time the lamp went dark.

Aside from the new lights, Alan also spent time sorting out the keepers. As with many in the early lighthouse service, they had been appointed on a temporary basis as and when necessary. Alan made the first steps towards establishing proper terms of employment, revising the lists of duties and lobbying the Commissioners for improved salaries and pensions. He also appointed a full-time lighthouse missionary to travel round all the NLB stations, ministering to the keepers and reporting back to George Street. With most of the remote lights, it was almost impossible to allow keepers to go to church on Sundays, and, as at sea, the usual practice was to have a service at the light conducted by the principal keeper. There were other, less spiritual preoccupations as well.

During 1846 and 1847, the potato blight which had already devastated Ireland, blew its way to the Highlands. In total, over 100,000 people were left destitute in the north-west, many of whom either emigrated to the New World or travelled to the cities in search of work. The disaster was exacerbated by the Clearances, which had reached their peak in the 1830s. There was little the Stevensons as individuals could do to alleviate the situation, but by 1847, matters had reached such desperate levels that the Commissioners willingly employed '50 starving labourers and all those dependent on them' in building a road to Eilean Glas light. The road was not strictly necessary, as an existing dirt track already provided an adequate link, but Alan saw the necessity of providing whatever employment he could. 'At first,' he reported to the Commissioners, 'many poor people came 9 miles daily to their work and returned at night to their huts, having only one meal a day; but temporary Barracks of Old Timber have been provided and meal has been supplied to them in part payment of their wages for which they are truly grateful.' When the famine reached epidemic levels in Caithness, Alan also began building roads at Nosshead and Dunnet Head. In the end, far fewer people died as a result of the famine in Scotland than in Ireland, due mainly to the reluctant but effective efforts of the landlords, the Church and the government relief co-ordinator, Sir Edward Pine Coffin.

Alan's workload prevented him from completing the last tasks at Skerryvore – the fitting of Soleil's lenses and the completion of the harbour at Hynish. Since David was now occupied with private business, Alan sent Tom in his stead. By 1844 Tom was twenty-six and supposedly fully independent. He had completed his apprenticeship five years previously and had already supervised the construction of at least one light alone. But he had proved a very different pupil from either of his brothers. Years of work for his father had made him disciplined, but he lacked Alan's driven creativity. Unlike David, he had a tendency to ramble off on his own schemes and interests, and what he gained in spontaneity he lost in focus. Even after five years of full qualification, he continued to lean on Robert, panicked by too much responsibility and prone to sour outbursts when left to his own devices. As he wrote petulantly to David in September 1842, while supposedly in charge of the construction of Little Ross light, 'I think I have written 5 or 6 letters to my Father and had a long letter from Elgin the other day without an answer to any of my queries ... I really *must insist* upon my correspondence being answered in a business like way as it places me in a stupid-awkward like box.' Robert, in his turn, still fussed over Tom, vetting his work and encouraging him with threats and promises. In Robert's eyes, Tom remained the worrisome youngest child, more cosseted than his brothers and less self-sufficient. When Tom wrote to his father asking plaintively for an assistant to help with some dredging work, Robert replied with a rare flash of humour that 'we must take care not to covet our neighbour's manservant.' Tom, it seemed, had tipped the balance with his father from interdependence to dependence.

Apart from occasional panics at his abandonment, Tom enjoyed his time at Skerryvore, though not, admittedly, for the pleasure in producing a well-made sluice system. He had been there once already in 1841, when the tower was fifty feet high. Unfortunately, it was not the light that had made an impression on him, but the seven-hour journey back to Hynish during which the boat crew had collapsed from seasickness. Three years later, he brought with him several workmen and a notebook which he filled with jottings and opinions. Most of the time, the work bored or exasperated him, since he had none of Alan's ability to ignore the outside world and thus treated most of his stay on Tiree as a form of grudging exile from Edinburgh. Confined to the island, he took no particular interest in the progress of the works and considered much of it a form of hard labour to be endured rather than enjoyed. He lacked the inclination to understand his workmates fully and usually pre-ferred the intricacies of building materials to the politics of individual relationships. On one occasion, when the scaffolding on the harbour wall collapsed, pitching several of the workmen into the bay below, he recorded laconically that 'such a scene of confusion presented itself as might have well made me laugh had not there been doubt as to the safety of those who were precipitated. Some were swimming while others were scram-bling among the floating logs. There was no damage done how-ever excepting a few scratches.'

On 31 May, in the same unconcerned tones, he noted that he and Dr Campbell, the resident Hynish doctor, had been out for a ride in the evening. 'I observed that he became gradually more and more incoherent in his conversation till at last he dropt off his horse. At first I thought it must have proceeded from effects of *vinum generosum* but as he was quite well when we came out I saw it was not that. It arose from his having eaten about an ounce of morphia lozenges during our ride. This is a very curious case. He was delirious during all that night and was unwell for several days after.' Poor Dr Campbell cannot have found Tom's reaction very reassuring. At moments it must have seemed as if he could cheerfully have poisoned half the workforce without his boss batting an eyelid. The only matter that did rouse Tom's attention were proceedings at the local kirk. Tiree, it seemed, was in danger of going over to the Free Church. Tom, who held the same conservative views as his father, found the dissenters distasteful. 'This week's work,' he wrote crossly, 'has given birth to a new sect of which it has been always thought there were already too many.' He could not see any problem with 'her Majesty's Ministers', and regarded the dispute as an aimless, vain squabble over 'peculiar dogma(s) of their own'. He worked himself into a righteous froth over the 'infidels, doubters and inquirers after a true religion'. His own belief was later to become a point of contention between him and Louis. For the moment, however, he confined himself to mutterings and insults in his journal.

Despite Tom's apparent irascibility, he did get something useful from his time on the island. His main entertainment was not the works or the workmen, but a series of obscure – and, to observers, absurd – experiments on waves. After complaining about the inconveniences of Tiree's roads and the monotony of the dredging works, he had taken to watching the sea. The size, weight, shape and impact of waves fascinated him. 'Nothing can be more beautiful,' he noted happily, 'than to see the harmonious intersections of these waves. The waves last night were by the wave pole [measuring height] made out to be 6ft high but they afterwards rose to a great height as our harbour works can testify.' He began filling his notebooks full of complex lists of different formations: retro-waves, right-angled waves and recoiling waves. He began tests with a home-made dynamometer, a primitive machine for measuring the weight and impact of water against a fixed object. The dynamometer was a box attached to a mechanism shaped roughly like a train's buffer, which, when fixed below the waterline, would retract according to the force of the water pushed against it. Tom fixed one meter on Skerryvore and the other on a rock near Hynish and took regular readings of the differing pressures. In every case, he noted, the impact of the waves was always greater on the reef than on the island.

Tom took vast pleasure in his discoveries, and later provided the lightkeepers on the Bell Rock and Skerryvore with logbooks in which, to their evident perplexity, they were instructed to record regular measurements. His fascination was later to pay useful dividends, though his relations were unimpressed by Tom's apparent capacity for paddling endlessly at the seaside and calling it work. As his son Louis later wrote in *Records of a Family of Engineers*, 'He would pass hours on the beach, brooding over the waves, counting them, noting their least deflection, noting when they broke. On Tweedside, or by Lyne or Manor, we have spent together whole afternoons; to me, at the time, extremely wearisome; to him, as I am now sorry to think, bitterly mortifying. The river was to me a pretty and various spectacle; I could not see – I could not be made to see – it otherwise. To my father it was a chequer-board of lively forces, which he traced from pool to shallow with minute appreciation and enduring interest ... It was to me like school in holidays; but to him, until I had worn him out with my invincible triviality, a delight.' Robert was more cryptic but equally damning. By the side of a squiggly, ink-blotted sketch of wave-formations in Tom's Skerryvore notebook, Robert wrote 'Perfectly Unintelligible'.

Once the harbour, dock and sluicing system at Hynish were finished and the lens fitted, Tom was free to take his experiments elsewhere. In December 1843, a notice was posted in the Scottish newspapers warning interested parties of a new light on the Skerryvore rocks to be displayed from the beginning of February

1844, 'and every Night thereafter, from sunset to sunrise'. Alan also drew up a careful specification of the light for the use of all shipping in the area. 'The Skerryvore light will be known to Mariners as a Revolving Light,' he wrote, 'producing a Bright Flash once every minute. The lantern, which is open all round, is elevated 150 feet above the level of the sea. In clear weather the flashes of the Light will be seen at the distance of six leagues, and at lesser distances according to the state of the atmosphere; and to a near observer, in favourable circumstances, the light will not wholly disappear between flashes.' Alan managed to make one hurried trip out to Skerryvore during his annual light-house tour. He was delighted with the lens and the completion of the works. 'We see the light [of Skerryvore] most magnificently from the hill,' he wrote. 'It is full of spiculae and is fully more splendid than Inchkeith although 3 times more distant.' The barracks which had caused Alan so much trouble remained until 1846 but had become so rickety that it was eventually removed. The whole enterprise, from the landing of the first stones to the fitting of the final provisions, had cost £90,268.

In Edinburgh, work continued at its usual pace. Most of Robert's work for the Convention of Scottish Burghs had been handed over to David, who was now responsible for the furniture of Stevenson work: river-dredging schemes, harbour construction, canal business. David was a thorough, painstaking worker, well-suited to the details of bread-and-butter engineering. Unlike Tom, who was more interested in his own enthusiasms, David stuck to the mainstream path of consultative engineering. He had also become fascinated by controlling and harnessing water forces, but regarded it in more straightforward terms than his younger brother as a test of science over nature. He became expert at gauging the effects of tides on coastal works, methods of diverting shoals or dredging estuaries to allow deeper-hulled boats through to port. More than any of the other Stevensons, David was beginning to define the course of marine engineering. During his time, it became less the adventurer's sport than it had been in his father's youth, and more the prosaic discipline it was considered by the end of the nineteenth century. David was, admittedly, not an adventurer in his father's mould. He liked thoroughness, convention, procedure and systems. Out on his own, he relished order, not the pleasures of risk. He was to become, in the best and the worst of senses, the most professional of the Lighthouse Stevensons. But if he lacked his father's magpie habits, he had many of Robert's steadier qualities. David was an engineer to the bone, who saw everything – geography, politics, history – through the prism of architecture and construction.

During 1836 and 1837, David had gone on a lengthy tour of English works, and then on a fact-finding trip to America. The journal he kept provides a glimpse of a man now indivisible from his work. On the trip from Liverpool to New York, for instance, he took pains to note the size, weight and towing speed of the tug pulling them out of harbour and spent most of the Atlantic crossing making notes on the air and water temperatures, latitude, longitude and wind speed. Few other observations intruded, other than the information that a young orphan boy had been found stowed away in the hold. The boy, wrote David noncommittally, 'gave no definite reason for his leaving for New York. He was tied up to the shrouds according to sea fashion – he was however soon released and set to sweep the decks &c by the captain. At 12 noon today the temperature of the air was 42° faht. and the water 49°.' Likewise, the first thing David noted on arrival in New York was not the glittering images of a foreign city, but a careful record of American dock- and harbour-construction techniques. His tour included Philadelphia, Washington, Mississippi, Ohio and the Great Lakes. Amongst David's impressions were an index of good, tolerable and poor US inns, the dimensions of American steamships, and several tetchy observations on the grubbiness of American trains. He remained an ambivalent tourist. He liked American engineering but found Americans themselves bizarre. As Robert reported to a colleague, 'He says he never was so thankful to God than when he got through, or over, the Alenghenny Mountains to Lake Erie. It is well, he says, to go on business, all he has seen is interesting, but to go into the Western States for pleasure is out of the question.' At least David's lists of specifications intrigued his father. He saw, noted Robert approvingly, 'steamers and other vessels of 600 or 700 tons!' The trip took eight months in total, and on his return through Europe he confessed that he would like to find a publisher for his impressions. David did eventually produce an account of his American wanderings which has since become a useful guide for scientific historians, if not a particularly compelling read.

David might not have recorded them, but other matters besides engineering were beginning to preoccupy all the Stevenson brothers. By 1848, all four of Robert's children remaining in Scotland had married: David to Elizabeth Mackay in 1840; Alan to Margaret Scott Jones in 1844, and Tom to Margaret Isabella Balfour in 1848. Jane had been married to Dr Adam Warden since 1828, and, to Robert's delight, had produced seven children, one of whom took over Jane's role as his companion and secretary. David had courted Elizabeth for a while, finding in her a kind-hearted counterpoint to the stiffness of Stevenson family life. By 1855, Elizabeth had given birth to eight children. Six survived and two, David Alan and Charles, later became the third generation of Lighthouse Stevensons. Tom's marriage to Margaret, the youngest daughter of the Reverend Balfour of Colinton, was also a fusion of opposites. She countered his moodiness with warmth and was cheery enough to deal with his occasional fits of melancholia. Her arrival also acted as a balance to Robert's influence. All four children moved away after their marriages but remained within the New Town.

Alan's marriage followed the pattern of the rest of his life. He had met Margaret Scott Jones in 1833, while on a working visit to the Welsh lights. Margaret, then twenty-one, was the daughter of the local landowner who disapproved of Alan on both financial and social grounds. He was, after all, the son of a bourgeois man who – regardless of his reputation, his successes and his stability – retained the sheen of new money and suspect effort. Moreover, Robert's philosophy had ensured that each of his children should earn their own way, just as he had. At the point when he met Margaret, Alan was still completing his training and had yet to be taken on as a full partner in the family firm. Most of the Stevenson apprentices paid Robert for their training; in his sons' cases, he waived the charge but did not give them a full salary until they qualified. Even when they did, it was only £150 a year, low even by the standards of the time. Alan was broke, in other words, and all the Cicero in the world would not convince Margaret's father he was suitable for his daughter. The two were banned from contact with each other, but managed to maintain a furtive correspondence. Much of Alan's communication came in the form of poetry, which, despite all Robert's frettings, he still wrote in secret. He prayed, he wrote in a poem of 1834, that she might 'never know the chill / Of cherish'd hope, requited ill.' In 1836, he was writing of the unnamed, 'yonder beach, / Where first our love was plighted; / Oh! meet me, lady, with the love / Distrust has never blighted.' In 1844, eleven years after he first met her, Alan married Margaret. He was thirty-seven, she was thirty-two.

The two settled in Windsor Street, not far off from Baxter's Place. By 1851, the couple had produced four children, two of whom, Robert Alan Mowbray (Bob) and Katherine Elizabeth Alan, were later to become lifelong companions for Louis. With the birth of his children, his new professional status and Skerryvore behind him, Alan appeared finally to have reached a point of contentment in his life. But just as Alan's life finally seemed to have reached its domestic and professional zenith, fate intervened. His workload had doubled again with his promotion to

Chief Engineer, and the endless dank journeys around the coast of Scotland were beginning to tell on his health. The annual tour had now become an unwieldy juggernaut of preparation, logistics and materials. Back in the early days, with only a dozen or so lights to supervise, Robert had found the journey a comparatively easy matter of landing supplies and checking on the keepers, but by the mid 1840s, there were around thirty-five lights scattered round every headland and islet of Scotland's outermost points. For one man to supervise the details of every single one was almost too much. For one man with precarious health making the journey several times a year in all weather it was the kind of load that only a Stevenson would have been stubborn enough to attempt.

In 1844, there were the first indications that the burden was beginning to take its effect on his strength. 'I am still grievously afflicted by Drowsiness,' he complained at Hynish in July while inspecting Tom's work at Skerryvore. By August, he was forced to row thirty-six miles in an open boat to reach Barra Head light, since there was not enough wind to sail. A few days later he tried to reach Calf of Man in a gale. For two days, he tried unsuccessfully to persuade the local boatmen to take him out, and on the third, 'I walk to the boat where I am nearly lost in the surf in embarking, having fallen back but am caught up by one of the boatmen.' In July of 1847, while on the annual voyage with the Commissioners, he noted shakily, 'I am ill with Rheumatism.' By the 25th, in Oban, he wrote, 'the night is v bad.' His handwriting became increasingly patchy and erratic over the next couple of days. 'In Lochindaal all day with gale,' he noted on Wednesday the 28th. 'I am v. ill.' Next day, 'leave Lochindaal at 3. I am so ill that the Commissioners kindly send me on to Ardrossan direct and I get home.' Two months later, he tried to complete the trip, but noted at Graemsay that 'I land and work in the rain although my suffering is extreme.' Eventually Tom had to complete the inspection tour on his brother's behalf.

In April 1848, having apparently recovered, Alan set out again, checking stores, lecturing keepers and reading sermons while storm-stayed in Scapa Flow. In August, he landed below Cape Wrath, 'and climb up the cliff below the lighthouse', an ascent of over 200 feet up sheer rock. 'A path is required,' he noted laconically at the top, 'for the Cattle to keep them from the cliffs over which they fall.' In 1849, he began the circuit again. 'An awful day,' he jotted in April. 'A hurricane from the NNE and thick snow all day. Lie in bed all day as we cannot stir and I have a heavy cold from yesterday's work.' At Barra Head, he found all the keepers complaining about their health, for which Alan prescribed a dose of reading: 'I promise to send the keepers some books; Newton, Leighton's *Poems*, Meickle's *Solitude Sweetened*.' By the beginning of May, having battled with gales, recalcitrant keepers and his own exhaustion, he noted only that 'I am very tired and ill.'

Alan's own account of his illness was erratic. In addition to an endless series of colds, fevers and bouts of flu, he had what was variously described as lumbago, rheumatism, paraplegia and an unidentified aching in his joints that felled him for days on end. Judging from the patches of good health he still enjoyed, the disease was not a fast-wasting one like motor neurone disease, nor the conventional debilitations of rheumatoid arthritis. From Alan's paralysing symptoms, and from his erratic periods of remission, it seems that the most likely cause was multiple sclerosis. MS, which attacks the myelin sheaths protecting the nerve fibres of the central nervous system, blocking and eventually destroying nerve endings in the brain and spinal cord, is now well known, but was not identified as a separate disease until 1868. Its progress is gradual and insidious, creeping through the body, blurring the vision and weakening the limbs as the nerve endings slowly cannibalise themselves. In the later stages of the disease it can become completely paralysing, and often results in a form of dementia. 'The patient,' records one medical directory bleakly, 'may also suffer from paralysis resembling the effect of a stroke, then eventually fall into a sometimes lengthy coma before dying.' Even now, treatment is based on staving off the symptoms; MS is still considered an incurable disease. Perhaps the cruellest aspect of it is the hope; the periods when movement and sight return, only to be withdrawn again by a further decline. Alan's work – slithering round rocks, clambering up cliffs, cold, damp and exhausted, could only worsen a process which had begun long before.

At first, he dealt with the illness by returning home to Edinburgh and seeking treatment from his brother-in-law Dr Adam Warden. With marriage and the duties of the NLB, he had lost the chance to bolt abroad to warmer, easier climates, and so, when the pains became too bad to allow him even to write, he would seek solace or a cure at one of the English spas. He began to scale down his workload a little. He did almost nothing now for the family business and concentrated as much as possible on the less strenuous business of the lights, testing lenses and dealing with paperwork. His brothers picked up some of the slack, but, as the disease crept onwards, Alan could do less and less.

In 1850, Alan's decline was matched by a further blow. On 12 July, a month or so after his seventy-eighth birthday, Robert Stevenson died. His wife, Jean Stevenson, had died four years previously and with her death Robert aged rapidly. For all his frailty, however, he remained active almost until the day of his death. On 28 June, he was still attempting to prepare a fine copy of his memoir of the voyage with Sir Walter Scott, despite his family's pleas for him to stop and rest. Louis, who was born four months after Robert's death, perhaps understood his unmet grandfather best of all. In one of his many memoirs, he recorded Robert's last days:

> In 1850, my grandfather began to fail early in the year, and chafed for the period of the annual voyage which was his medicine and delight. In vain his sons dissuaded him from the adventure. The day approached, the

obstinate old gentleman was found in his room furtively packing a portmanteau, and the truth had to be told him ere he would desist: that he was stricken with a malignant malady, and before the yacht should have completed her circuit of the lights, must have himself started on a more distant cruise. My father has more than once told me of the scene with emotion. The old man was intrepid; he had faced death before with a firm countenance; and I do not suppose he was much dashed at the nearness of our common destiny. But there was something else that would cut him to the quick: the loss of the cruise, the end of all his cruising; the knowledge that he looked his last on Sumburgh, and the wild crags of Skye, and the Sound of Mull with the praise of which his letters were so often occupied; that he was never again to hear the surf break in Clashcarnock; never again to see lighthouse after lighthouse (all younger than himself and the more part of his own device) open in the hour of the dusk their flowers of fire, or the topaz and the ruby interchange on the summit of the Bell Rock. To a life of so much activity and danger, a life's work of so much interest and essential beauty, here came the long farewell.

The Commissioners, gathered together the day after his death, recorded a solemn obituary to 'this zealous, faithful and able officer', and gave their condolences to the remaining Stevensons on the loss 'of one who was most estimable and exemplary in all the relations of social and domestic life.' Their regret was doubtless genuine, but the record unwittingly emphasises the irony of Robert's life. Despite all his inventions and endeavours he remained eternally the servant, never the master. Almost all of his work had been done with the explicit intention of providing something for others – a public service, a monument to industry, a method of saving other men's lives

– and Robert took the benefits and the drawbacks of that philosophy. The Bell Rock had given him the recognition he had always wanted, but he did not, unlike Rennie, Telford and many of his engineering contemporaries, move on to more glamorous projects. His work remained functional, the stuff of public highways or municipal schemes rather than the architecture on which Victorianism prided itself. If people noticed his bridges, harbours, docks and drainage schemes, they did so only because they were well built and stood the test of time. Even the lighthouses were unlikely to invite anything except the fleeting gratitude of a passing mariner. The fame went to others, and the Bell Rock remained his masterpiece. He was forced to take his satisfactions from the more elusive charms of his trade, in hard work and enduring strength. Initially, he had been content with his role as public servant, but, as his peers outstripped him and the admiring notices went to other, grander names, the lack of recognition nagged away at him. Others, he felt, did less work for better rewards. As he wrote to a friend many years after the completion of the Bell Rock, 'When I look back on my long experience as a Marine Engineer I often blame myself for not standing more forward.' Perhaps there is also a certain piquancy in the fact that his last task had been to write up his record of a journey with another, more famous man. It was a final try at celebrity, this time by association.

When Robert died, some of the qualities that he had brought to his work died with him. His methods – learned more through the practice than theory, discovered painstakingly through trials and errors, through improvising where necessary and adapting where possible – had been superseded as much by his own efforts as through time. Engineering was no longer the amateur trade that it had been when Robert began, it was now a respectable profession. When Robert had begun the Bell Rock, there were so few experts in marine architecture it was possible for him to communicate personally with every one of them. By the time of his death, many universities had established separate engineering courses and young men were pouring into apprenticeship schemes similar to the one Robert himself had introduced. His sons, though pioneers within their own specific sciences, were not and did not need to be that all-round 'man o' pairts' that Robert had been. Even the lighthouses, an esoteric sideline compared to the traditional marine fare of bridges, harbours and dockworks, were now a known and studied quantity. Smeaton, Robert and Alan had, between them, brought the rock lighthouse to a degree of perfection that few others could, or would need to, match. Tom and David, for their parts, wisely found different territory on which they could make their marks.

Meanwhile, Alan struggled on for two more years before succumbing to his own illness. On 9 February 1853, the Commissioners received his resignation. In an emotional letter read out before the Board, he admitted that his departure would not come as a surprise to them. He was now 'under the necessity of begging to be permanently relieved from duties which I grieve to say there is now no prospect of my being able to resume. I cannot take this step so important to the interests of myself and family, without much pain; nor can I relinquish without many serious regrets, an Occupation with which I have for so many Years been familiar ... The Circumstances and duration of my servitude and emoluments are so perfectly known that I will not enter into any detail on the subject.' His letter was accompanied by two letters from doctors, certifying that Mr Alan Stevenson was afflicted with an unusually serious form of paraplegia. He 'has ceased to make such progress towards the recovery of his health ... he can never resume the arduous duties of his office.' The Commissioners accepted his resignation with 'deep regret', considered the question of his pension, and awarded him half his annual salary. A few months previously, crippled by pain and hopelessly disillusioned by the path his life had taken, Alan had inscribed a poem on the title page of his young son's bible.

Read in this blessed Book, my gentle boy;
Learn that thy heart is utterly defiled . . .
This day five years thou numberest; and I
Write on a bed of anguish. O my son,
Seek thy Creator, in thine early youth;
Value thy soul above the world, and shun
The sinner's way; oh! seek the way of truth.
Oft have we knelt together, gentle boy,
And prayed the Holy Ghost to give us power
To see God reconciled, through Christ, with joy;
Nought else, but Christ brings peace in sorrow's hour.

God and poetry remained Alan's only consolations. He had always been devout but in his retirement he became almost maddened by his own guilt. God, it seemed to Alan, was teaching him through suffering. He saw his paralysis not as an arbitrary tragedy, but as God's retribution for past sins. Every time the pains lifted, he had moved a little way to paying his penance; every time they came back, God was punishing him. He became tortured both in mind and body, and as the disease began to take its insidious toll on his brain, he lost his ability to see himself as anything other than a sinner eternally punished. At one stage, he went through an agony of conscience over his insistence that the Skerryvore men should work on the Sabbath. In 1854, a decade after the light was completed, he wrote to each one of them apologising. 'I blame myself,' he confessed, 'as sole cause of this violation of the Law of God; and I state this with great pain, duly humbling myself before Him, not only as an individual transgressor, but as a deceiver of others . . . Being now awakened to the great wickedness of such conduct, I feel bound to confess my sin, not only to God, against whom it was committed, but before you, who were or might have been deceived by me into a sinful unrighteousness.' The God that Scotland believes in has always been unusually retributive, quick to punish and slow to forgive, making, particularly in His more zealous, Calvinistic manifestations, a particular speciality of guilt. After his retirement, Alan seems to have worshipped a uniquely Scottish God.

With Robert's death and Alan's decline, the business of the lighthouses was left entirely in the hands of David and Tom. Alan Brebner, one of Robert's apprentices, whose father had worked on the Bell Rock, was appointed as deputy to the brothers, while David took over the role of chief engineer in Alan's stead. His first task was a project so awkward it was to become known as his Skerryvore. By 1853, war with Russia was looking increasingly likely, and France and Britain were mustering troops. By the time of its conclusion two years later, the Crimean War had earned perpetual notoriety for the incompetence of its military leaders. The British aims in the war – to curtail Russian expansionism in the east and to cauterise the threat from France – were obscured by a catalogue of stupidities culminating in the Charge of the Light Brigade. At the time, however, Westminster was preoccupied with despatching a naval fleet to blockade the White Sea ports at Archangel and Murmansk. To do so meant a journey northwards through Scottish waters around the top of Norway and Lapland to Russian waters.

Following a tetchy correspondence between Trinity House, the Board of Trade and the Commissioners, it was concluded that the fleet would need at least two new lights around the north and east of Shetland to guide them through the stormy waters towards Scandinavia. David was therefore despatched northwards to establish sites and survey the ground. He set off at the beginning of March 1854, pursued by filthy weather. Twice, he tried to land on the northernmost Shetland isle, Unst, and twice he failed. He was not a man in the habit of overstatement, but his account of the trip and his findings dispensed with all the Stevensons' usual rules of dispassionate summary. The journey was evidently so terrifying that David had spent much of it fearing for his life.

On returning to Edinburgh, he was adamant that the seas around the Shetland coast made building a lighthouse in the area impossible. 'It is not practicable,' he reported bluntly, 'to erect and maintain a lighthouse upon these rocks' It was too dangerous, he concluded, too expensive, and any ship that took that route was mad anyway. The passage between Orkney and Shetland was already well lit and it would be much more sensible for the convoys to make their way through the more placid route. 'The maxim that it is better *not to exhibit a light at all* than to run the risk of any thing approaching to failure . . . must be regarded not only as a *safe*, but as a *necessary* rule in Lighthouse Engineering,' he wrote emphatically. He had no choice. The Admiralty were insistent that the north coast must have lights, and no amount of prevarication by the NLB and its engineer was going to distract them from that purpose. The Admiralty, it must be noted, did not actually inspect the coast themselves, though they were uncomfortably aware of a disaster in 1811 during which three naval battleships (the 98-gun *St George* and two 74-gun ships) were shipwrecked in a North Sea storm while returning from the Baltic. Two thousand lives were lost, double that of the British naval losses during the Battle of Trafalgar.

David had no choice but to capitulate. In an extraordinary situation, he was forced to use extraordinary means. Accordingly, he recommended establishing a temporary light at Whalsey on the east, and at Lambaness, with the option of building a further light on North Unst. The Admiralty promptly overrode David's reservations, insisting that North Unst was the most important site, irrespective of the dangers involved. Matters were not helped by a deputation of Elder Brethren from Trinity House, who arrived to view the site on a day of such pearly calm that the journey constituted more of a pleasure trip than a working tour. They were, they reported blithely, of the opinion that a light on North Unst was 'quite practicable'.

David, mumbling wrathful comments about the Elder Brethren's 'fair-weather' jaunts, was limbering up for full-scale battle against Trinity House when the Admiralty sternly told both parties to forget their quarrels and get on with building the light.

One glance at a map of Scotland reveals how much David's reservations were justified. The Admiralty were demanding a light on the last piece of land between Shetland and the Arctic Circle. Unst is closer to the Faroe Islands than it is to Edinburgh. By the lines of latitude, it is farther north than St Petersburg or Greenland's southern cape. The Admiralty wanted a lighthouse built on a rock just off the northernmost tip of the island, a place known with quaint charm as Muckle Flugga. The rock itself, the 'Great Precipice' is less charming. It forms part of a large reef protruding vertically out of the sea like a range of miniature Alps. The main rock is not a flat shelf as the Bell Rock and Skerryvore had been, but a steep triangle rising sheer from the water. The seas in the area were so atrocious that it was commonplace during winter for unbroken waves to sweep right over the 200-foot summit of the rock, a height 50 feet above the top of Nelson's Column. While David was conducting his initial survey he made note of a six-ton block of stone which had been torn from its moorings eighty feet above sea level and hurled into the sea below. This, he reported drily, 'clearly proves that on these coasts we have elements to encounter of no ordinary nature.' In addition to the monstrous seas, there were also the attendant threats of newly revitalised press gangs patrolling the area, the difficulties for the lighthouse ship in bringing supplies for a light, and the problems with staffing it when built. Matters were not helped by the Shetlanders' suspicion of the Edinburgh men. Once informed that war with Russia had broken out, the locals refused to help with the light, and appeared to believe that David and his assistants were Russian themselves.

The Commissioners reluctantly decided that a temporary

light was to be built on the summit of the Flugga and provided with separate accommodation for the keepers. The stone for the light was to be quarried from the rock itself, and the rest of the materials shipped from Edinburgh. On 31 July, the lighthouse ship sailed from Leith with a cargo of men and a hundred tons of materials. The job of resident engineer for the works had been delegated to Robert's old assistant Alan Brebner, who was to be relieved in winter by Charles Barclay, with David as overall supervisor. Work progressed quickly, though David was keen to point out the difficulties involved. 'When it is considered that the whole of the materials and stores consisting of Water, Cement, Lime, Coal, Iron Work, Glass, Provisions &c, and weighing upwards of 100 tons had to be landed at an exposed rock, and carried up to the top on the backs of labourers, it will be seen that the exertions of Mr Brebner who conducted the Works . . . [have] been in the highest degree praiseworthy.' Even in relatively mild weather, the ship could not moor too close to the rock, so (as at Skerryvore) everything to be landed had to be strung together with rope, swung overboard and hauled up the temporary wooden ladders to the summit. In all, 120 tons of materials were dragged bodily up the rock by the workmen and fitted into place. By 11 October, three months from the start of works, the temporary light was ready.

David's misgivings remained. On 2 December, a violent storm began brewing. Within two hours it reached hurricane force. The supervisor on the rock, Charles Barclay, reported to the Commissioners that 'the sea was all like smoke as far as we could see and the noise which the wind made on the roof of our House and on the tower was like thunder . . . the water came pouring in upon us, so that we had to drive it out (as much as possible) with a mop and brush &c before the next spray came.' The next morning, Barclay, the workmen and the three terrified keepers realised that the storm had removed a large section of the dyke, shifted three immense water casks several yards and broken one of the lantern panes.

Little more than a week later, another force-12 gale hit the rock. This time, reported Principal Keeper Marchbank, by now almost incoherent with weariness and fright, 'such was the violence of the Wind and Spray that it carried up Earth and Stones with such Violence against the Lantern of the Tower that it broke one of the panes of Glass in the Lantern we immediately got in the Storm pane and it broke one of the panes of it and at the same time it Broke Two of our Lamp Glasses and nearly extinguished our Light such was the Violence of the Wind and spray that we all thought they would carry everything before them ... it was 4 O'Clock in the morning before any of us got to bed Mr Charles Barclay and his Workmen say they never heard any thing like it in their time it was blowing so strong and the Spray so heavy.' This time, the coal house was blown down, the dyke breached in several places, and the water had come unbroken through the dwelling house. During the lulls, Barclay and the men had worked frantically to repair the damage and strengthen the tower. It made little difference. On 31 December, another storm stole forty feet of the dyke, six of the water casks and a chunk of the ladder. The barracks was now so thoroughly soaked that all the bedding and supplies had to be destroyed. 'We had not a dry part about all the premises,' wrote Marchbank unhappily. 'We have been very uncomfortable for the last Month both in the Lightroom and Dwelling house when we had not a dry part to sit down in nor even a dry Bed to rest upon at night.' For the seas to rise unbroken over a 200-foot rock presented difficulties that no one, not even the Stevensons, could have planned for.

David was so concerned for the keepers' safety that he felt bound to report to the Board of Trade that 'life is in jeopardy'. If the Commissioners had their way, he noted pointedly, everyone would have been removed from the rock and the light discontinued for the winter. For once, Trinity House backed him up, remarking that those who lived on the rocks were the best judges of conditions, and conceding the 'serious danger to which the lives of the Light Keepers would be exposed by a continued residence on the rock'. They did, nevertheless, feel that Muckle Flugga was still the 'most proper site' on which to have a lighthouse, whether temporary or permanent. The Board of Trade and the Admiralty also agreed that, despite the hazards to human life, the risks would be far greater if the light was removed. In the face of implacable opposition to the withdrawal of Muckle Flugga, David's solution was to suggest strengthening the light, making it permanent instead of temporary. If the men could not be brought off, at least they could be made safer than at present.

Establishing the permanent light meant a reversion to more traditional methods. Workmen's cottages had to be built near the site, one of the lighthouse ships taken out of service at Leith and stationed permanently on Shetland, and staff, from storekeepers to boat crew, appointed. As David noted glumly in his report on the new works, 'The place is wholly without local resources and every thing even to a ropes end must be conveyed to the Spot no assistance is to be derived from the natives excepting in boating and all masons, quarriers &c must be taken from this [the allocated funds] and high Wages paid to induce them to go.' This time, David had decided to use brick instead of the customary granite to build the light, since brick was less troublesome to transport, easier to haul up the rock, and dispensed with the need for awkward quarrying. The rock itself was so flawed that it splintered into fragments during extraction, and without a deep cladding of plasterwork would have let water come pouring into the tower. 'The whole of the upper part of the Rock consists of rotten Shivery material,' wrote David, confessing that, though the use of brick in such stressful conditions was 'an untried experiment in marine engineering,' it was almost unavoidable. The walls, he emphasised, were to be built three and a half feet thick and the whole tower constructed anew. The weather remained atrocious. On one occasion, a wall of water crashed over the summit of the rock, stripping the tower of materials and the men of possessions. Much of the time, the danger of both wind and waves was such that the workmen had to crawl out of doors on their knees for fear of being heaved off the rock and out to sea.

By the end of 1857, the new light, constructed under Alan Brebner's supervision, was finished. From that day to this, it has never let in a drop of water. David returned with relief to the ordinary lighthouse business, and the quarrel with Trinity House and the Board of Trade faded away. But matters remained volatile. London's insistence that the Commissioners should barter over every detail of the light threatened to paralyse the works completely at times. At one point, the Commissioners, incensed that their judgement was being overridden by committee men down south, threatened to resign their duties and hand the whole business of the Scottish lights over to someone else. David, they argued, was not only a competent engineer, but knew his territory better than any Trinity House man could ever do. If London thought it knew better, then London could see how it fared with Shetlandic gales and impossible deadlines. Faced with the awesome prospect of a dozen mutinous Scots sheriffs, the Board of Trade retreated a little, applied a few emollient words and conceded that the Commissioners probably did know something after all. It was only one of an eternal series of quarrels between Edinburgh and London. At its heart remained the old sore of English meddling in Scottish affairs.

There were, however, more important matters than wars and rocks. The Stevensons' children were growing up, flocking through the Edinburgh schools and chattering off to North Berwick during the holidays. Alan was aching away the remainder of his life in expensive English spa towns, paralysed by poverty and self-doubt as much as by his own enforced inactivity. His son, meanwhile, was finding consolation in the company of his cousins. Domestic life was taking precedence again; the next generation of prospective Lighthouse Stevensons were appearing. And Tom's son Louis was being groomed for the life of an engineer.

EIGHT

Dhu Heartach

The story of Robert Louis Stevenson's childhood has become almost as well known as his fiction: the long sick nights drifting through the Land of Counterpane, the endless attempts to find medicine or comfort, the ministrations of his nurse, Alison Cunningham (known as Cummy), terrifying him with tales of brimstone and Covenanters. 'I have three powerful impressions of my childhood,' he wrote later, 'my sufferings when I was sick, my delights in convalescence at my grandfather's manse of Colinton, near Edinburgh, and the unnatural activity of my mind after I was in bed at night.' His weak health has usually been attributed to a history of bronchial problems on his mother's side. His Balfour grandfather was weak-chested, and most biographers see him as the source of Louis's sickliness. But matters were not much better on the Stevenson side. Robert's rigorous good health was the exception, not the rule. Alan, like Louis, had suffered from ill-health since childhood, and both David and Tom were prone to bouts of illness. Louis therefore inherited a double dose of frailties, as well as a family perpetually checking its own pulse. Louis's condition only worsened his father's existing capacity for hypochondria. During Louis's childhood, Tom developed an endless series of phantom illnesses in sympathy with his son.

Tom and Maggie both doted on Louis. Nursing and worrying about him became a full-time occupation during most of his childhood. The rest of his time was spent with Cummy, who stoked his imagination on a diet of Calvinist history and adored him with all the passion of a thwarted mother. The atmosphere of the house in Howard Place took on its own feverish quality with Tom downstairs writing passionate tracts on waves or non-conformism, Maggie fretting over her husband and her child, and Cummy soothing Louis with another tale of retribution. When Louis was a little older, the family would board a boat for France or Italy to take the waters and find healthier air. Cummy would accompany them, and on one journey in 1863 she kept a diary famed for its pungent disapproval of everything foreign. She complained about the devilish papist practices of the Continentals, and found most of Italy hopelessly inadequate compared to Edinburgh. Rome, she wrote, 'is about the size of Edinburgh, though not nearly so bonny as Auld Reekie'. Catholic practices she found worst of all. 'No wonder the Frenchmen think our quaint Scottish Sabbath dull,' she wrote smugly, 'for here is everything to please the unconverted heart of man – worldly pleasure of every kind, operas too!'

When she wasn't hurling herself into churches and complaining about foreign food, Cummy kept a record of the ailments that plagued the Stevensons. Mrs S. had a nasty cough, Mr S. had troublesome piles, Mrs S. was afflicted by headaches, Cummy herself had seasickness and 'peculiar feelings', while Louis ran the gamut of everything from fevers to bleedings. The sick little band of Scots wandered from place to place, perpetually chasing better health. But if illness made them a gloomy party, it also made them a close one. Tom's concern for his son showed itself in an unexpected sweetness of spirit. When Louis woke screaming from nightmares, only his father could reassure him. As Louis later acknowledged, Tom would 'rise from his own bed and sit by mine, full of childish talk and reproducing aimless conversations with the guard or the driver of a mail coach, until he had my mind disengaged from the causes of my panic.' Tom had his own strange preoccupations, and there was a deep seam of understanding between father and son which was later to form the basis of many of their ferocious arguments.

Despite Tom's dotings, there was still work to be done. David was still nominally chief engineer to the NLB, but the two brothers nearly always worked in tandem, David tending towards organisation and Tom towards details. Tom's interest in the shape of waves was also beginning to form itself into a well-respected expertise. After completing his work at Hynish, he despatched log-books and dynamometers to the keepers at Skerryvore and the Bell Rock, instructing them to take regular note of wind speeds, the height of the spray and the maximum water pressure per square foot. The keepers were evidently nonplussed by Tom's 'wave engines' and letters of instruction, but were shrewd enough to keep their own counsel. Tom also collected anecdotal accounts. The oldest fisherman in Aberdeen, he discovered, had stated that the waves were always highest when the wind was from the south-east; '10 of the oldest and most intelligent fishermen' in Argyllshire stated unequivocally that the worst storms were from the south-west; and one of the keepers at Calf Point wrote in describing an unusually fierce sea lifting a ten-ton block of stone from the landing slip and shifting it several feet to landward. There was also some documentary evidence of storms so powerful they overturned normal physical laws. During the seventeenth century, a merchant ship was wrecked in a hurricane off the west coast while carrying £750,000-worth of gold coins and ingots. When the ship's cargo was salvaged, it was found that the structural ironwork was embedded with one of the ingots and several sovereigns. If waves could solder gold into iron, it was evident that there were forces at work that no amount of clever engineering could account for.

In all, Tom made 267 experiments over two years, which he finally published in the *Journal* of the Royal Society of Edinburgh. Atlantic swells, he noted, were almost always heavier than those from the North Sea. The pounding that Skerryvore took in a heavy gale was almost twice what the Bell Rock endured. In a later essay, he cited the experience at Muckle

Flugga and at Whalsey Skerries, where six-ton blocks of stone had been lifted clean from their foundations seventy feet above sea level. Tom was fascinated by the sea's random malignity. At one point he described standing on the cliff at Whalsey examining the 'indications of a violent, destructive agency which seemed to have been lately at work upon the hard rock . . . The only visible agent was the ocean, the unruffled surface of which appeared far below the place where I stood . . . Here, then, was a phenomenon so remarkable as almost to stagger belief – a mass of 5.5 tons not only moved at a spot which is 72 feet above high water spring tides, but actually quarried from its position *in situ*.' In published papers, Tom kept his findings dry, but in private jottings he took his experiments as proof of the existence of something far greater than straightforward science. As Louis noted, 'storms were his sworn adversaries,' something he took personally. They provided him with a battleground, a place where man and nature confronted each other. The Stevensons had made their name and trade out of subduing the elements, and Tom understood that confrontation with particular astuteness. It worried him that his knowledge of the sea was, and would always remain, strictly finite, and that, despite all the advances of the last century, the sea still had the power to shock. Standing above a placid ocean, watching the displaced masses of rock and gauging the immensity of the forces that had moved them both frightened and intrigued him. His family, it seemed, could build lighthouses till the end of time, constructed with all the tenacity that stone and physics permitted, and yet the sea could still wipe out every fragment within seconds.

Faced with such an incomprehensible force, Tom's usual reaction was to take notes. He filled more books with scribblings, skittering from an admission that measuring waves was 'excessively difficult and unsatisfactory', to passionate notes on the Crucifixion and 'heathen writers'. 'All waves are to a greater or lesser extent waves of translation,' he mused, next to 'Observations on a Remarkable Formation of Cloud at the Isle of

Skye', and the assertion that, 'The nonconformist is often one of the most overbearing of modern pests.' His beliefs were always vehement, his interests always fanatic. He was a black-and-white man at war with a life of greys. Small wonder that even Louis struggled to reach the furthest points of his father's character. As he pointed out, Tom was not a born scientist, yet he had chosen to pick an intellectual fight with the ocean itself. 'He was a man of a somewhat antique strain,' mused Louis many years later, 'with a blended sternness and softness that was wholly Scottish and at first somewhat bewildering; with a profound essential melancholy of disposition and (what often accompanies it) the most humorous geniality in company; shrewed and childish; passionately attached, passionately preju-diced; a man of many extremes, many faults of temper, and no very stable foothold for himself among life's troubles.' He had inherited Robert's dogmatism but not his self-confidence. The arguments between Tom and Louis became more than just an illustration of a father's hopes for his only son; they were also about Tom's own precarious hold on the world.

When back in Edinburgh, Louis's greatest friend was Alan's son Bob. He was the child for whom Alan had written his despairing poem and was already developing into a character, 'more unfitted for the world', as Louis put it, 'than an angel fresh from heaven'. Their games of pirates and adventures were necessary pleasures for both of them. Bob wanted a refuge from the dark atmosphere of Alan's decline at home, and Louis pro-vided it. The prospect of their father trapped in a fading body and tortured by Godly remorse was scarcely welcoming to his children. When possible, Bob escaped to Edinburgh to stay with Louis, or to make sandcastles down in North Berwick. They rode ponies. Bob had 'Hell', his sister Katherine had 'Heaven' and Louis rode 'Purgatory'. Sometimes, David's chil-dren would accompany them. His two sons, Charles and David (known as David A., or DAS), had been brought up in prim Edinburgh style, and lacked the wildness of their cousins. Louis, in fact, found the two disgracefully smug and always referred to them both as 'the good little boys' until his adult conscience caught up with him. Bob, on the other hand, was an all-rounder like his father, just as happy with mathematics as he was with art. He had, as Louis described it, 'the most indefagitable, feverish mind I have ever known; he had acquired a smattering of almost every knowledge and art; he would surprise you by his playing, his painting, his writing, his knowledge of philosophy, and above all by a sort of vague, disconnected and totally inexplicable erudition.' In later years, he went to Cambridge and then became a writer and critic on art, though no particular subject pinned him down for long. He, like Louis, revolted against his father's beliefs. Whereas Alan had suppressed the best of himself to become an engineer, Bob challenged everything openly. As with Louis, the biggest breach erupted over religion. Bob found his father's devotion impossible to understand, and in later life became a vehement dissenter.

Bob's youth and rebellion coincided with Alan's final decline. After his retirement, he and the family had moved first to Portobello and then to the Fifeshire village of St Cyrus, within sight of the Bell Rock. As the MS worsened, Alan had to make frequent trips to the English spa towns to help alleviate the pains, and his NLB pension was dwindling fast. There was no money to send Bob to the High School, no money for holidays or spoilings. The little Alan could do – scientific papers, snippets of poetry, memoirs of his father – did not pay well, and the increasing periods of illness meant that he was unable to take regular work teaching or translating. He did manage to contribute a large entry on lighthouses to Chamber's *Encyclopaedia*, and, more for the sake of occupation than reward, began translating the 'Ten Hymns of Synesius' from the Greek. Synesius had been ambassador to the Emperor Arcadius Augustus, and had written a series of 'sublime and grand' verses as well as several essays including a 'Treatise on Dreams' and a tract 'In Praise of Baldness'. Coleridge, whom Alan knew through Wordsworth, claimed that he had himself translated eight of the Hymns by the age of fifteen; Alan claimed modestly that without the poet's lead, he would 'not otherwise have ventured to try my feeble and unskillful hand on the work'. He wrote a brief introduction to the translations, mentioning that 'it pleased God in 1852 to disable me, by a severe nervous affliction, for my duties, as engineer to the Board of Northern Lighthouses; and I took to beguiling my great suffering by trying to versify the whole Ten Hymns of Synesius. During many an hour, the employment helped to soothe my pains.' With the translations were a number of his own poems, all heavily influenced by Wordsworth. Most of them contained a deep seam of religious anguish. One poem, entitled '*On Memory as an Agent of Retributive Justice*,' emphasised his preoccupation with punishment and atonement.

> In that dreadful day,
> When the last trumpet's wondrous note shall sound,
> Rending the spheres, and piercing the dull tomb,
> The wheel of each man's destiny shall roll
> Backwards, unfolding all his inner life;
> From his last breath to childhood's earliest sin,
> Which led him first from God, all shall be told.

Finally, in 1865, Alan Stevenson, the quiet pioneer, died. The Commissioners recorded an unusually heartfelt tribute, noting 'their deep and abiding regret for the loss of a man . . . whose genuine piety, kind heart and high intellect made him beloved.' By any standards, he had been an extraordinary man, fluent in six languages, a classical scholar, the friend and champion of the era's most influential poets, a pioneer of optical technology, and the architect of one of the most exceptional structures ever built. Skerryvore, the lonely tower in a lonely sea, became his lasting monument, 'immovable, immortal, eminent', as Louis put it. The time he spent on that dark rock in the mid-Atlantic, battling with wind and wave, gave him something close to happiness. The years afterwards, sick and persecuted, seemed a hard penance. Throughout his life, from the first submission to Robert to the final submission to illness, there remains a whisper of the might-have-been. Sometimes, as at Skerryvore or pondering lenses with the Fresnels, Alan gives the impression of a man fulfilled in his work. At other times, he gives the sense of someone bent out of true by more forceful wills than his. Some measure of his character is still contained within Skerryvore itself, built brilliantly, but also built at great personal cost.

With Alan now gone and Tom preoccupied, it was David and Elizabeth who took on Robert's role as Stevenson grandees. They lived the fine refined New Town life, gave parties and supported the post-Disruption Church of Scotland. David, not a questioner by nature, brought up his sons with the intention that they should inherit the lights. He also passed on many of his pragmatic qualities. Both Charles and DAS picked up a sense of pedantry in details and thoroughness in life. David did not believe that imagination was a necessary quality in an engineer. One had only to glance at his sons' feckless cousins, Bob and Louis, to see the proof that too much intellectual freedom was bad for the soul and the finances.

In 1855, David made the crucial choice between the lights and the general engineering work. Sole reliance on the NLB, he decided, was a dead end, too specialised and too insecure to support both brothers for long. In an address to the Commissioners, he pointed out that his workload had increased several hundredfold. Three new lights were to be built in 1854 alone, not counting the works on Muckle Flugga, the paperwork had risen from 150 orders and logs to over 1,000 documents for each light, and the bureaucracy surrounding the annual accounts had doubled. 'When I had the honour to be nominated Successor to My Brother the late Engineer to the Board, the impression was that the business for the future would be less extensive than it had been,' he wrote. 'New arrangements have however since been made by which the duties connected with the Ordinary management instead of being diminished have been greatly increased.' With such a load in addition to his own troubled health, 'which has never been very robust since I had an illness about 3 years ago of more than 2 years duration, constrain me to intimate that I am unable to undertake the duties of Engineer to the Board on the present footing, an intimation which I make with all the reluctance and hesitation which naturally attaches to retiring from an office which has so long been held by members of my family.' Nominally, David and Tom should have been sharing the work equally, but, given the differences in their characters, it was inevitable that much of it had been delegated. David, as always, had a clearer organisational mind. When one side of the business faltered, it was he who fussed it back to health. Tom was usually too busy with his own esoteric schemes to devote his full attention to all aspects of the business and, though the two were complementary partners, the burden of the NLB and the family business had inevitably devolved on David.

David suggested that the Commissioners might consider appointing Tom and himself jointly as general engineers to the Board, ridding them of the need to deal with unnecessary paperwork. If the two were to work as partners, concerning themselves only with essential business, then the burden of the lights could be better balanced. Forthwith, work would not be delegated to one person alone, but to two separate departments. One, supervised by the Stevensons, would design and plan new works and advise on developments such as changes to the lenses or fuel sources. The other would deal with the ordinary business of the lights, such as inspection tours, preparing accounts, buying stores and managing the keepers. The division, he argued, would strip the engineer's department of non-engineering work, since much of the work they were currently responsible for did not require any form of technical expertise. As he pointed out, the Board of Trade's insistence on picking over the NLB accounts with a fine tooth comb had created enough work to support another full-time employee. Finally, he noted that 'should the Board see fit to entrust the New Works to the charge of my brother and myself . . . we should use every endeavour to urge them on as vigorously as possible.' Other contractors, he noted delicately, were simply not as good as the Stevensons. He and Tom knew the lights like no one else. When other engineers had been hired, their performance 'lead[s] us to anticipate that we could conduct the works much more expeditiously than perhaps under any other arrangement which the Board could make'. After due consideration, the Commissioners approved the idea and appointed a separate General Manager to help with the bureaucracy of the service.

From then on, therefore, David and Tom were free to return to the purer details of planning new lights. Between 1854 and 1880, the two were responsible for the construction of twenty-nine lights around the coast, from Butt of Lewis and MacArthur's Head to Ushenish, Monach, Skurdyness and Lochindaal. Commissions were also beginning to come in from abroad. Foreign governments, encouraged by the reputation of the NLB and the gathering fame of the Stevensons' achievements, looked to the Scottish lights as their templates. Tom and David were asked to supply lights first for the Indian government and then for Japan, Canada and Singapore. The commissions were lucrative but daunting. Until then, they had been accustomed to working with purely local conditions – freezing waters, Atlantic ferocity, hurricane-force gales, everything that a sullen northern climate could fling at them. But they did not have experience of the problems that came with designing lights for earthquake zones and tropical temperatures. In a high Indian summer, for instance, the lights could overheat, the lantern glass crack and the lead melt. Japan, on the other hand, had a similar climate to Scotland, but had to take account of fault lines and earthquakes. The East also had its share of other problems. The waters of China and Malaya supported a thriving population of pirates and wreckers who responded predictably to the threat to their trade. In several cases, the Malayan lights had to be encased in so much protective lead that the cost of defending them far exceeded the cost of constructing them. The engineers working on the Malacca Straits lights had to be armed at all times to defend themselves against the wreckers, and even the keepers were forced to carry guns. In practice, however, the years of experience in Scotland had provided the Stevensons with almost all the knowledge they needed to build a lighthouse almost anywhere in the world. Dovetailing stone blocks provided flexibility during tremors, thickened glass withstood monsoon rains, and iron lights built on stilts (similar to the rocket-shaped barracks at the Bell Rock and Skerryvore) could be used for sandbanks or wreckers' territory.

Tom, meanwhile, was preoccupied with lenses. His dioptric holophotal light was an adaptation of Fresnel's earlier work, but instead of using a set of fanned mirrors above the lens to drive the light outwards (as Alan had chosen for Skerryvore), it enclosed the light in a prismatic glass casing. The popular image of a lighthouse lens (an immense convex apparatus with thick myopic portholes in the centre, topped and tailed with angled prisms) was, in part, Tom's lasting contribution to optics. The new dioptric holophotal light (holophotal meaning 'whole light' in Greek) finally dispensed with the need for reflectors, and cleared the way for the immense lenses of the late nineteenth century, tall as two men and over four tons in weight. Coupled with David's adjustments to the angle and reach of the lenses, the inventions meant the Scottish lights were the most accurate and powerful in the world, capable of being turned, angled, coloured and eclipsed with unprecedented ease. Lights could now be adapted for almost any purpose. The beam could be strengthened or reduced, forced out in one direction only, rotated faster or slower, or given a different 'character' with the use of extra or coloured lenses. The strength of the new lenses reached up to a million candlepower. Smeaton's twenty-four tallow candles at the Eddystone had only reached around thirty candlepower.

Experiments were also being conducted into new fuel sources. In 1860, a correspondence began between the NLB and Trinity House's special scientific adviser, Michael Faraday, over the relative merits of 'magneto-electric' lighting. Faraday, now considered the founding father of electric power, was already famed for his lectures at the Royal Institution in London conducted against a backdrop of theatrical explosions. Trinity House, keen to discover whether his inventions could be applied to lighthouses, agreed to allow him to conduct a series of tests on electric lighting in 1858. Faraday, with the help of a scientist named Holmes, selected the South Foreland light for his experiments and set up a mechanism emitting around 60,000 candle-power with a colza-oil light nearby as comparison. As he reported to the Elder Brethren, the light shone strongly, but 'has a tendency to sudden and spontaneous extinction . . . the liability causes an anxious watchfulness on the part of the attendant, who does not descend to the guardroom, but is constrained to stop in the lanthorn continually.'

Despite the difficulties in keeping the light on and the expense of fitting the equipment, Faraday noted triumphantly that the electric light was far brighter than the old oil lamps. 'On going out to the hills round the Lighthouse, the beauty of the Light was wonderful,' he reported. 'At a mile off, the apparent streams of Light issuing from the Lantern were twice as long as those from the Lower Lighthouse and apparently three or four times as bright . . . The tops of the hills, the churches and the houses illuminated by it were striking in their effect upon the eye.' He conceded that fitting new electric lights would be expensive, that the keepers would need to be retrained to deal with the light and, modestly, that the light was often so bright that it might give mariners a false impression of the distance from ship to light. But his experiments were positive enough for him to wholeheartedly recommend using electricity in many more of the English lights. 'The light produced,' he concluded, 'is powerful beyond any other that I have yet seen so applied, and in principle may be accumulated to any degree; its regularity in the Lantern is great, its management easy and its care there may be confided to attentive keepers of the ordinary degree of intellect and knowledge.'

The NLB, however, was more cautious. As Faraday conceded, the new magneto-electric light was unlikely to be suitable for remote places, and would be expensive to fit and maintain. Tom took a trip to England in 1865 to inspect Faraday's lights, and returned both impressed and concerned. He found the power of the new light as brilliant as advertised, considering approvingly that 'in all cases superior power constitutes in reality superior safety.' But he was wary of its habit of blacking out and pointed out that, in its present form, electric light was hopeless for remote lights like the Bell Rock or Muckle Flugga. In the short term, the dispute about its merits was subsumed by the usual trouble over lack of funds. It wasn't until 1883 that the Board of Trade passed enough money to allow the Stevensons to set up a trial light at the Isle of May. In the meantime, the lights got by on paraffin, gas and goodwill. (Gas proved unexpectedly awkward; when fitted at Leith and connected to the local supply, it was discovered that it paled or died completely every time the local shopkeepers went to work.) Faraday, meanwhile, kept up a desultory but good-natured correspondence with the Commissioners, and appeared at one stage to dispense advice to keepers on filtering water at the lights in case of cholera.

Back at home, Louis was reluctantly coming to terms with engineering. His father's threats and pleadings had finally brought about much the same conclusion as they had a generation previously. Louis was to spend his winters in the south studying theory, and his summers in the north supervising different Stevenson projects. For three long summers, Louis tried dutifully to adjust to his new profession. Admittedly, his first posting to Fife in 1868 did not much encourage his enthusiasm. 'I am utterly sick of this grey, grim, sea-beaten hole,' he wrote to his mother. 'I have a little cold in my head, which makes my eyes sore; and you can't tell how utterly sick I am, and how anxious to get back among trees and flowers and some thing less meaningless than this bleak fertility.' Occasionally, he made it obvious how far he was from adjusting to his new profession. 'What is the weight of a square foot of salt water?' he asked his father plaintively, 'and how many lbs are there to a ton?' With his mother, he was more honest. 'Tell Papa that his boat-builders are the most illiterate writers with whom I have ever had any dealing,' he wrote to Maggie.

Louis's apprenticeship coincided with the greatest challenge of his father's career. By 1857, Tom had concluded that something must be done to light the infamous Torran reef, twelve miles off the Ross of Mull and thirty-three miles south-east of Skerryvore. The main rock, the 'black and dismal' Dhu Heartach (now spelt Dubh Artach) 'is an egg-shaped mass of black trap, rising thirty feet above high water mark', according to Louis. 'The full Atlantic swell beats upon it without hindrance, and the tides sweep round it like a mill-race. This rock is only the first outpost of a great black brotherhood – the Torran reef that lies behind, between which and the shore the Iona Steamers have to pick their way on their return to Oban. The tourist on this trip can see upwards of three miles of ocean thickly sown with these fatal rocks, the sea breaking white and heavy over some and others showing their dark heads threateningly above the water.' Years later, he used the reef as the 'stoneyard' on which David Balfour and Alan Breck were shipwrecked in *Kidnapped*.

The Commissioners began to collect petitions from captains and seamasters who used the area. All of them argued strongly in favour of a light; Captain Ticktack of the *Palmer*, who had been carrying a cargo of logwood from Jamaica to Liverpool, was wrecked on Seil Island. He had, he declared, 'lost Seaman, Wife and Child as well as Ship', and considered that a light on Dhu Heartach would have saved him 'all his subsequent sorrows and losses'. Captain Bedford of the Royal Navy reported the fate of the *JP Wheeler*, bound for Glasgow from America. On 31 December 1865, the captain had attempted the tricky passage between the Ross of Mull and the Torran Rocks, while the Mate busied himself with necessary repairs. The captain 'was sitting at the side of his bed when he heard a heavy sea break upon the deck – his idea was that the Vessel was in the tideway of the North Channel – running out he met the Mate coming to report that the Ship was among the breakers. It was too true, for all around him the sea was breaking as high as his topmast head. He went up the Mizzen rigging – thought he could discover a smooth, set more sail (which he expected every moment would go to ribbons) and succeeded in extracting his Ship from the Torrens Rocks.' Tom himself was in no doubt about the need for the light. During the annual inspection of 1854 he had tried to make a landing on the rock, but even on a sunny day with a placid sea, it proved to be 'impracticable'. Between 1800 and 1854, he noted, thirty ships had been wrecked on the reef, with upwards of fifty lives lost, and several ships, including the *JP Wheeler*, seriously disabled. 'We have no hesitation,' he concluded in a joint report to the Commissioners, 'in reporting that the erection of a Lighthouse upon it would be a work of no ordinary magnitude.'

It would, he suggested, be a project as complex as Alan's venture two decades earlier. The Torran reef was not an isolated hazard, as Skerryvore and the Bell Rock had been. It was a long, scattered archipelago of mainlands and islands stuck slap in the centre of a major shipping channel. The reef extended for around ten miles in total, and the surrounding area was cluttered with protruding stumps of rock, some lying underwater, some just above. The difficulty was not in finding a place on which to build, but the opposite problem. Tom could have lit the reef from end to end and still not been confident it was marked well enough. He therefore needed to pick his site with care, to ensure that the light would be seen from all approaches. In particular, it had to be visible to mariners sailing up the thin corridor to the Firth of Lorn and the lower end of the Sea of the Hebrides, which was prone to powerful currents. The number of outlying islands in the area and the driving Atlantic swell from the west pushed the unwary either towards the reef or towards the land nearby. Any ship that drifted into the area might avoid one rock only to be knocked to pieces against another. As Tom pointed out, the whole place was snared with natural traps as impassable as anything Alan or David had ever contended with, and nothing but the smallest row-boat could navigate the area with any hope of safety.

The site he eventually selected for a tower was a humpbacked lump of stone rising about 40 feet above high water mark and extending for 240 feet in length. The rock itself was a smooth black crest of seaworn basalt, slippery to walk on, but at least more pliable than the cramped gneiss of Skerryvore. The drawback was the difficulty in getting any purchase on the rock, since Dhu Heartach presented one sheer black mass rising from the waters without a landing place or sheltered creek from which to land materials or moor supply boats. After the intervention of Lloyds of London – one of the parties involved in governmental decisions on British waters – the Commissioners conceded the need for a light, and authorised Tom to start work. Tom's designs for the tower followed Alan's Skerryvore template in miniaturised form. The light was to be built 101 feet above foundation level to the parabolic curve (Skerryvore, by comparison, being 138 feet and hyperbolic), it would have seven floors and would be quarried from similar granite to that used by Alan. A shore station would be established on Earraid, a tiny isle connected to the Ross of Mull by a tide-washed spit of sand. Earraid was admittedly over fifteen miles away from the reef, but it was the nearest land available that provided shelter, a vantage point and a reasonable mooring place. As with the Bell Rock and Skerryvore, Tom intended to build an iron barracks in which the workmen could stay during the progress of the works. The whole undertaking, Tom calculated, would cost around £57,000, roughly the same as Alan's initial estimate for Skerryvore.

The shore works, at least, progressed with gratifying speed. The stones were quarried, cut and shaped, and the workmen's cottages completed in record time. Fifty men or so had set up permanent camp by the shore, under the charge of Robert's old apprentice Alan Brebner. By the time Louis arrived in 1870, the settlement was well established. On the shore, he noted,

> there was now a pier of stone, there were rows of sheds, railways, travelling-cranes, a street of cottages, an iron house for the resident engineer, wooden bothies for the men, a stage where the courses of the tower were put together experimentally, and behind the settlement a great gash in the hillside where granite was quarried. In the bay, the steamer lay at her moorings. All day long there hung about the place the music of clinking tools; and even in the dead of night, the watchman carried his lantern to and fro in the dark settlement and could light the pipe of any midnight muser. It was, above all, strange to see Earraid on the Sunday, when the sound of the tools ceased and there fell a crystal quiet. All about the green compound men would be sauntering in their Sunday's best, walking with those lax joints of the reposing toiler, thoughtfully smoking, talking small, as if in honour of the stillness or hearkening to the wailing of the gulls. And it was strange to see our Sabbath services, held, as they were, in one of the bothies, with Mr Brebner reading at a table, and the congregation perched about in the double tier of sleeping bunks; and to hear the singing of the psalms, 'the chapters,' the inevitable Spurgeon's sermon, and the old, eloquent lighthouse prayer.

Building work on the tower itself began in April 1867, although for several months, the weather remained so foul that little was achieved in the first year. The foundations to the workmen's barracks were fixed to the rock, but here was no time even to bore holes for the diagonal struts before high seas made further digging impossible. When work started again in April 1868, the weather was so bad that the workmen were unable to make more than a snatched hour's visit to the rock until late June. Even then, Tom glumly reported, they had only managed two days' work in June, thirteen in July, and ten in August. Those few August days were as much by default as by design. Thirteen of the workmen under Brebner's supervision had seized one of the few fine-weather opportunities and sailed out to the rock. Instead of returning to Earraid at dusk, Brebner took a gamble on the good conditions, and chose to stay on at the barracks for the night. That evening, caught unawares while they were making the last checks, a six-day storm broke over their heads. The workmen barely had time to collect their tools and run for shelter before they found themselves stuck in the middle of an Atlantic hurricane. There was no question of returning to the boats. The only option was to sit and wait it out. All fourteen of the men were stuck in their refuge for almost a week, 'during the greater part of which time the sea broke so heavily over the Rock as to prevent all work,' Tom reported to the Commissioners, 'and during the height of the Storm the spray rose high above the Barrack, and the sea struck very heavily on the flooring of the lower apartment.' At one moment, the sea swept in through the iron trapdoor at the base, swirled around the workmen and then disappeared through the hatch, taking with it most of their remaining food supplies. Most of the time, the workmen had been too terrified to sleep, and could only hunker down in a corner of the room for warmth. When they did finally return to Earraid, they were met by an apoplectic David, frantic with worry for their safety. Brebner, cowed, stuck rigidly to caution from then on.

The incident did at least prove exactly how intransigent Dhu Heartach could be. As Tom explained to the Commissioners, the barracks was sixty feet above high water, and should in theory have been safely out of the sea's reach. But, as the last two years' working seasons had revealed, the reef was proving to be just as difficult as Alan had found the one at Skerryvore. During the next two years, work on the rock remained sluggish, constantly interrupted by foul weather and heavy swells. The most temperamental part of the process – the landing of the moulded stones for the light – caused an endless series of problems and delays. Winching two-ton blocks of granite up the glass-smooth surface of the rock without a proper landing stage proved just as dangerous as David had found it in Shetland's far more extreme conditions.

Tom appeared occasionally from Edinburgh to chivvy or instruct. During the 1869 season, he took Louis on the lighthouse inspection tour. They sailed round the east coast lights, up to Scapa Flow and then to Muckle Flugga, but Louis seemed far more interested in the scenery than he was in the lights. At Earraid the following year he seemed no happier. Just as at Wick and Anstruther, he hated the work. He was not made for formulae and tonnages, and spent most of his time chafing to be away. In later life, he regarded his time as an apprentice with some affection, though he had in reality stayed only a short while and disappeared back to Edinburgh as soon as his father would allow him.

The works progressed fitfully in Tom's absence. Gales blew and the weather beat down, and he was forced reluctantly to concede that the working season had to be restricted to just three months in mid-summer. The lantern was eventually fitted and lit in 1872; the whole project had taken a year longer to complete than Skerryvore. It had been an interesting exercise. No matter how practised the Stevensons became in building lights, Dhu Heartach had demonstrated that there would never be a definitive template for all situations. The reef was only a few miles from the great black battlefield of Skerryvore, but the peculiarities of its shape and weather had forced an entirely different working pattern from Alan's experience. Two years after the completion of the light, a further unexpected hazard appeared. In 1874, the principal keeper wrote to David, informing him that they had been sitting in the lighthouse kitchen when 'we heard a rumbling noise, followed by a tremendous motion which lasted for about two seconds . . . a fresh gale from WSW was blowing at the time, but there was no sea striking the rock to cause the concussion; in fact there was less sea than had been for some days previously. When a heavy sea strikes the tower, it has quite a different effect and cannot be mistaken for anything else . . . I can offer no suggestions as to the cause, unless it proceeded from a slight shock or earthquake; the rumbling noise and tremendous motion indicated such.' This was not a problem the Stevensons usually encountered, though Scotland does occasionally experience minor tremors. Fortunately, the dovetailed stone construction of all the lights made them more than flexible enough to compensate for small rattlings.

With the completion of Dhu Heartach, Tom turned his attention back to other business. But if he had needed further proof of the waywardness of nature, events at Wick provided it. Harbour works had long provided a staple diet for the Stevenson firm. These were mundane bread-and-butter to them, but essential for the small fishing villages they served. Tom was responsible for the construction of a breakwater at Wick on the instruction of the local Town Council. It was an ordinary commission, executed with the ordinary exactitude of all Stevenson work, and consisted of a long crooked arm of stone and masonry extending out from the harbour into which boats sailed for shelter from the winds. As was now customary with Stevenson projects, it was designed by Tom, but supervised by one of the junior partners in the firm. Tom would appear at intervals during the building work to pick over details and then move on. Louis was also sent there in 1868 as part of his training, and loathed the area with a memorable loathing. 'Wick itself,' he later recorded, 'is one of the meanest of man's towns, and situate certainly on the baldest of God's bays.' He remained indifferent to the harbour, and only showed an interest in proceedings when he was offered a trip down in a diving suit to watch the underwater builders at work. Lowered into floating twilight, surrounded by the immense foundations and the littered stones, Louis found the experience a better lesson in physics than any amount of book-study. He was also enough of his father's son to derive some pleasure from watching the sea. In a letter to his mother in September 1868, he mentioned standing by the new pierhead, observing the results of a winter gale.

> The end of the work displays gaps, cairns of ten ton blocks, stones torn from their places and turned right round. The damage above water is comparatively little: what there may be below, *on ne sait pas encore*. The roadway is torn away, cross-heads broken, planks tossed here and there, planks gnawn and mumbled as if a starved bear had been trying to eat them, planks with spales lifted from them as if they had been dressed with a ragged plane, one pile swaying to and fro clear of the bottom, the rails in one place sunk a foot at least. This was not a great storm, the waves were light and short. Yet when we were standing at the office, I felt the ground beneath me *quail* as a huge roller thundered on the work at the last year's cross-wall.

Tom, he supposed blithely, would have been less impressed by the artistry of the storm than by its effects. 'I can't look at it practically however: that will come I suppose like gray hair or coffin nails.'

When finally finished, long after Louis had departed for more promising places, the breakwater stood intact for four years until a spectacular storm in December 1872 destroyed the entire harbour, shifting one massive block of stone weighing 1,350 tons and folding the whole structure into the sea. Tom was devastated. In fact, his reaction was far more extreme than the incident warranted. But he had based his professional faith on studying the sea, learning its moods, its tempers and its breaking points, and the discovery that much of his life's work was founded on a miscalculation was almost unbearable. The early studies he had made of the force of waves were based on the movements of ten- or fifteen-ton blocks, not of something that weighed as much as the whole mass of the Bell Rock lighthouse. His reaction was initially incredulous, then defensive. He published papers complaining of the force of the elements the Stevensons contended with, photographs of immense waves smashing against the harbour walls, anything that might vindicate his position. Eventually, once the disputing was over, the breakwater was rebuilt, this time with a 2,600-ton foundation block in place. In 1877, another apocalyptic storm washed it away. Tom could do nothing but turn away in disgust.

On his return from Earraid, Louis had skipped off gladly down south before returning to Edinburgh to continue his training. In 1871, after delivering his paper to the Royal Scottish Society, Louis summoned his courage and confessed to his father he would never make an engineer. The compromise, advocacy, was scarcely more satisfactory, as Louis crept further and further towards the life of writing. 'You have rendered my whole life a failure,' Louis reports Tom shouting at him a year later. His decision also put great strain on Thomas's relations with the rest of his family. To justify his son's behaviour, he blamed Bob, which had led to several quarrels with Alan before he died, and to compensate for the disappointment of Louis's disappearance from engineering, he was forced to blame DAS and Charles for becoming all that their father had wanted them to be.

The disputes between Tom and his son were only an incidental backdrop to the lighthouse work. Back in the offices, now based, as they have been ever since, at 84 George Street in Edinburgh, commissions were coming in from both New Zealand and China. The Stevensons' existing expertise with the Indian lights made future work abroad more straightforward. Many requests consisted only in shipping off lenses and machinery without the need for training keepers or designing lights. Much of it was now undertaken by Alan Brebner, who had been made a full partner, and, increasingly, by DAS and Charles. Both Tom and David were beginning to suffer from bouts of ill health. In their case, it was not the slow wasting sickness that Alan endured, but the more prosaic effects of age and fine living. By 1881, David was forced to retire and handed on much of his work to his sons. Tom took over as senior engineer to the Board, but he too was prone to illness. Louis's decision not to take up the family business had proved a serious discouragement. Once David retired, Tom had less heart for the lighthouses. He did not work happily with David's sons, who seemed to him to represent a living reproach for his own son's waywardness. Much of the next few years was taken up with dynastic rivalry between Tom and the fourth generation of Lighthouse Stevensons. The fact that both DAS and Charles were proving to be excellent engineers was not something that Thomas wished particularly to dwell on.

The end, when it came, came fast. Four years after his retirement, David died. Two years later, in 1887, Thomas followed him. Despite their quarrels and their differences in character, the two brothers had made an exceptional team, with the absences in one complementing the strengths of the other. David had undoubtedly been the better suited of the two to engineering. Tom always remained a little out of balance with the formulae of day-to-day working practice. If such a thing exists, then David had been the perfect engineer, with cool judgement, a strong sense of vision and an intimate grasp of the interlocking worlds of architecture, design and science. The lighthouse that remains as his lasting testimony, Muckle Flugga, was built against his better judgement. It is a measure of his skill that the impossible lighthouse still stands almost 150 years later. And moreover it is through him that the third and fourth generations of Lighthouse Stevensons went on working the shores of Scotland long after he, Alan and Tom were dead. He and his sons were largely responsible for creating the NLB in its present format, streamlining the engineers' duties, defining the different tasks and completing the transition from an ad-hoc enterprise to a professional organisation. Much of his work was unspectacular. It was built to be useful, not to win prizes. But he more than any of Robert's sons ensured that the name and work of the Stevensons kept its shine.

Tom, as always, was an altogether more complex character. The indecisiveness of his youth, the later conservatism, the erratic enthusiasms and moods, the battles with men and oceans, the bitter dismissal of David's sons, the quarrels and reconciliations with Louis, all give a picture of a man forever fighting his own contradictions. In his own way, he was, like Alan, a thwarted man. If Robert had been a little more generous towards writing and a little less dogmatic in his ambitions, it seems probable that Tom might have made himself into an academic or a teacher. He was well suited to specialisation, he possessed a strong didactic streak, and he was so passionate about his own interests that he was often in danger of neglecting the ordinary Stevenson business. He too made his mark. Dhu Heartach was his monument and Wick his epitaph. Inevitably, however, it is not for lighthouses or lenses or harbours or good works that Tom has been recognised, it is for the double-edged honour of having sired Louis.

The deaths of Tom and David meant a sea-change in the NLB work. Suddenly, the engineers' department was no longer monopolised by Stevensons. Alan Brebner was now the senior partner, and work was increasingly contracted out to other firms. The work diversified from the usual business of building and maintaining lights to different refinements – foghorns, beacons, buoys and so on. The estrangement between Trinity House and the Commissioners also escalated. The Elder Brethren took a greater say in the workings of the NLB, vetoing new projects and arbitrating over those to be undertaken. Both the Commissioners and the Stevensons found their intervention exasperating, and relations sometimes deteriorated into open confrontation. As with Muckle Flugga, the Commissioners pointed out that conditions were different in Scotland and that local knowledge was essential. Increasing centralisation, they suggested, would only diminish the service the NLB provided. It was not always an argument that persuaded London. The skirmishes with the Elder Brethren and the Board of Trade (now the Department of Transport) continue to this day.

Yet despite the obstacles, the lights kept on being built. Fidra, Oxcar, Ailsa Craig, Fair Isle, Helliar Holm, Sule Skerry, Stroma, Tod Head, Noup Head, Tiumpan Head, Killantringan, Hyskeir, Trodday, Rubh Re – between 1885 and 1915, twenty-three new lights were constructed, from the Flannan Isles in the far north-west to Bass Rock opposite the summer beaches of North Berwick. The lighthouse service expanded and changed. The Stevensons took on other work and other lives, keepers came and went, fuels and lenses changed or dwindled. Through it all, the storms and wrecks still came, and the lights – colza or paraffin, electricity or gas, dioptric or holophotal – remained. Far away in Samoa, Louis was carried in a coffin to the top of a high hill. A separate industry developed to engineer his image and keep his own reputation burning. In the years after his death in 1894 he became a hero, the teller of tales, the perpetual exile, the children's author, the man of whom myths were made. He himself helped the process a little, but returned again and again to the place and the trade from which he had come. The rest of the family remained in Scotland, still connected to the Northern Lights in mind if not in deed. The last of the Lighthouse Stevensons, D. Alan Stevenson, died in 1971, two centuries, ninety-seven lighthouses and four generations after Robert first joined Thomas Smith. And the lights remained there at the edge, still turning.

NINE

The Keepers

All of Britain's lighthouses have now been automated. There is no longer any such thing as a lighthouse keeper. The profession has expired, forced into obsolescence by time and technology. Inside the towers, the brasswork and polish have been stripped away and in their place are banks of monitors and computer fittings. Tidy impersonal rows of electronic aids now turn the lights on and off, measure the daylight, calculate the windspeed and battle back the storms. The buildings remain, as plain and weatherbeaten as ever, but the keepers themselves have moved on.

The arguments for and against automation have long ago grown stale. On the one hand, there's the dismantling of 210 years of exceptional tradition, the death of a profession and the drawbacks of abandonment. On the other is the truth that at the tail end of the twentieth century it does not require three grown men to keep a light bulb. Some talk hopefully about the remanning of American and Scandinavian lights, but without real conviction. The tide of logic and change dictates that lightkeeping has become a redundant skill. Everyone agrees on the sadness; everyone agrees on the sense. Indeed, to many people, it is mildly astonishing that keeping persisted as long as it did, and its disappearance is no more than the death of a minor novelty.

The impulse that created the lights also seems dated at the end of the twentieth century. The sense of public honour that drove Robert and his sons might be useful in other contexts, but their ideology now seems anachronistic in the middle of a seascape populated by supertankers and nuclear submarines. High Victorianism, with its undertones of imperial arrogance and emotional humbug, has not stood up well to the test of time. But within the enclosed space between the first light and the last, there are still lessons to be learned or tales to be told. Inevitably, it is the keepers themselves who worked the lights long after the Stevensons had gone who provide the sharpest focus on the life of the lights.

In its heyday during the 1890s the NLB employed over 600 keepers in the 80 or so lights speckled around the coast, and kept a waiting list of over 200 hopeful applicants. Where possible, the NLB tried to recruit those with some maritime experience – ex-naval and merchant seamen or retired fishermen. In part, their previous jobs reduced the need for training, but the military discipline of most sailors' lives was also an asset. Even at the beginning of the service, the physical requirements were not hard. The NLB looked for candidates with a working knowledge of mechanics and regular habits. More important were the emotional qualifications: self-discipline, patience and a seen-it-all demeanour. Even in the mid 1990s, the NLB considered that 'A lightkeeper must be a man of parts . . . from his study of the sea, he will respect its immense power; he will be a handiman of varying proficiency, he will be a useful cook and a good companion. A lightkeeper will not make a fortune but the odds are that he will be at peace with himself and with the world.' For all its peculiarities, the job certainly had advantages. As the Commissioners put it in 1857, 'From the day that a Lightkeeper enters the service he possesses a certain income, a free furnished house of a description greatly superior to that occupied by the same class in ordinary life, all repairs and taxes paid, coal, candle, clothing, land for a couple of cows or an equivalent in money. He is provided for in case of age or infirmity; and, lastly, a provision is made for his widow or family in case of death. The only thing that can deprive a Lightkeeper of these secured benefits is his own misconduct.' Once you were a keeper, that was it; you moved steadily upwards from supernumary to assistant to principal, shifted once every few years and retired contentedly at age sixty with a civil service pension and the sea forever singing in your ear.

It was Robert Stevenson who set the standard for the keepers, and Robert's ghost who dictated their lives for the next 200 years. He introduced the military habit of mind that characterised the service, and a set of habits and rituals that persisted well into this century. One of the practices he introduced was to spring surprise visits on the keepers. The timetable of the annual inspection voyage was usually well-known, given that Robert and the Commissioners always sailed from Leith anticlockwise round the coast in mid-summer, but the programme of new works usually meant that he made several other journeys during the year. The first the keepers would know of his arrival was the lighthouse ship slipping around the edge of a nearby headland and Robert stamping up the pathway to the tower. Several of the keepers were caught comprised by his visits, either still asleep, or with undusted lanterns and unwashed crockery. When he did find something amiss Robert was a ruthless judge, with a bloodhound's nose for trouble or dust. After reprimanding the keeper at Kinnaird Head for his idle housekeeping, he confessed, 'It is the most painful thing that can occur for me to have a correspondence of this kind with any of the keepers, and when I come to the Light House instead of having the satisfaction to meet them with approbation ... it is distressing when one is obliged to put on an angry countenance and demeanour. But from such culpable negligence as you have shown, particularly of late, there is no avoiding it.' Slow deliveries, shoddy habits, cheap workmanship and idle practice all goaded Robert into an almost peevish rage. Even the weather, when it delayed or hampered his progress, was rebuked for its intransigence.

All the keepers were expected to be as virtuous in private as they were in public. As a profession, they were instructed to be 'sober and industrious, cleanly in their persons and linens, and orderly in their families'. Evidently it was no good hiring someone, however dutiful they might appear, if there was a risk of them being drunk, lazy or dishonest while on duty. 'I hold it as a fixed maxim,' Robert wrote to one disobedient keeper, 'which I have often put to the test of experience, that where a man, or a Family put on a slovenly appearance, in their houses, stairs and lanterns, I always find this; Reflectors, Burners, Windows and Light in general ill attended to; and therefore I must insist on cleanliness throughout.' By the plaintive tone of much of Robert's questioning – why had he not received reports from each keeper? Why had they not written as instructed? How was the new glass withstanding the weather? Were they keeping the wicks correctly? – it was evident that many of the keepers lay as low as possible to avoid the Stevenson whirlwind. At times, Robert's attitude took generous forms, such as his lobbying for better staff pensions, but it could also appear meddlesome, since Robert was not in the least bit squeamish about interfering in the private lives of his staff. 'I had a call the other day from Old Hunter, Cobbets Master,' he wrote slyly to Quintin Leigh, skipper of the lighthouse yacht in 1822, 'On mentioning to him that you had been complaining, he said in distinct terms that a *Wife* would entirely remove your complaints.'

All the Stevensons seemed to hold a dim view of the keepers' characters. Even Alan, usually the most tolerant of the family, reported to a Select Committee in 1845 that 'we have no information excepting from one or other of the keepers, and we generally find them very ready to give information against each other, for it is remarkable that they are generally on very bad terms; I know not how, but so it happens.' As Louis reported, no doubt encouraged by Thomas, the keepers 'usually pass their time by the pleasant human expedient of quarrelling, and sometimes, I am assured, not one of the three is on speaking terms with any other . . . The principal is dissatisfied with the assistant, or perhaps the assistant keeps pigeon, and the principal wants the water from the roof. Their wives and families are with them, living cheek by jowl. The children quarrel; Jockie hits Jimsie in the eye, and the mothers make haste to mingle in the dissention. Perhaps there is trouble about a broken dish; perhaps Mrs Assistant is more highly born that Mrs Principal and gives herself airs; and the men are drawn in and the servants presently follow.' As Louis noted, Robert had been shrewd enough to use this dissent to his own advantage. In one of his diaries he had noted that 'the lightkeepers, agreeing ill, keep one another to their duty.'

Those duties might not have been onerous, but they were specific. The keeper at Port Patrick light was given instructions in 1802 to trim the wicks of each oil lamp down to precisely three-sixteenths of an inch, and reminded that he was to 'clean the reflectors with linen or cotton rags, and then use a brush for cleaning the throat of the reflectors, and thereafter use the brush upon the surface of the Reflectors to complete the cleaning, that he clean the windows in the inside with linen or cotton rags, and on the outside to clear the glass of the dimness they are so apt to contract when the sea runs high with fresh water and a mop.' Furthermore, he was reminded 'that he keep always in remembrance that the lives and fortunes of numerous individuals often depend upon the exact performance of the duties of a lightkeeper.' Often, the keepers' wives helped out with the duties of the light, cleaning, housekeeping and maintaining the records when necessary. The majority of mainland lights allowed for families, with neatly divided cottages for the principal and the two assistants and ground for grazing a cow or sheep. The NLB never appointed female keepers, on the illiberal but understandable grounds that one woman bundled up with two men for four years on a place like Skerryvore was asking for trouble. On the rock lights – those such as the Bell Rock that were too remote or small to allow for more than the keepers themselves – the wives remained on shore nearby. The life of a lightkeeper's wife was similar in many respects to a fisherman's. She endured the same long waiting, the fright at storms and gales, the solitary motherhood, the stranger come home every few months to greet the bewildered children. If you married a keeper, you married the job, as several keepers' wives point out even now. Every few years or so, just as the children had settled into a new school, the transfer instructions would come through, and the family would move on, up north, to the isles, down to a city, out to the west.

Since their jobs therefore affected not just the keepers themselves but all those dependent on them, almost every aspect of their lives had to be vetted. As Robert discovered, there were not only the usual problems in maintenance and discipline, but unexpected difficulties to contend with as well. Dazzled birds came crashing through the lantern panes, keepers' wives needed maternity care, assistants sabotaged the principals' cows, the pigeon post got blown off course or taken by hawks. Robert's response was characteristic. For every new problem there was a new instruction. Books were to be dusted on the first Saturday of every month, parasols and umbrellas were banned from the light rooms, lenses were to be polished with chamois skin and vinegar every day. Any visitor 'in a state of intoxication' was banned from the lighthouse premises. Principals were to get ten tons of coal every year; assistants received eight tons. Principals 'must treat Assistants with courtesy and civility, but, at the same time, with firmness. Any act of disobedience, insolence or disrespect, on the part of an Assistant lightkeeper towards the Principal is forthwith to be reported to the Secretary.' Travelling arrangements, correct letter-writing procedure, reading matter, medication, church attendance, maternity care, every cranny of the keepers' existence was investigated and corrected by the Commissioners at some point during their tenure.

By the time of Alan's promotion to Chief Engineer, reading matter for each lighthouse had been established. Each light was to receive copies of useful instruction books and approved fiction. Wives were encouraged to study Francatelli's *Cookery for the Working Classes, How to Manage a Baby*, and *Miss Nightingale on Nursing*. Keepers with young children were to have the first and second *Books of Reading*, the *Rudiments of Knowledge*, the *Moral Class Book*, the *Introduction to the Sciences* and the *Geographical Primer*. The Commissioners also appointed a missionary to visit the more remote lights, both to ensure that the keepers had not taken up any hunnish practices and to provide some form of outside teaching. 'As regards the instruction of the younger members of the families in religious and secular knowledge, I made it a point to collect them together for this purpose at least twice each day,' wrote George Easton, the missionary between 1852 and 1893. 'In addition to Bible lessons, I taught them writing, reading, arithmetic, geography, grammar and English history (Scottish history was not generally taught to Scottish children until the 1960s). 'I know no class of men who have it so much in their power to remedy by personal exertion the unquestionable evils connected with their isolated positions as Lightkeepers; and I have never missed an opportunity of impressing this upon them.' Copies of the *Weekly Scotsman* and the *Illustrated London News* were also sent to each light, though, given the stormy weather and the unreliability of relief boats, they were usually well out of date by the time they arrived. Inevitably, the novels on offer usually included a generous dose of Sir Walter Scott.

Diet and medication were also well covered. By 1873, each keeper was allowed a daily ration of a pound of butcher's meat, a pound of bread, two ounces of oatmeal, barley, and butter, a quart of beer and as many fresh vegetables as could be grown at the light. The vegetable gardens were so jealously guarded that keepers were allowed to take any movable crops with them when they transferred to a new light, or to sell them on to the new keeper. All lights were also supplied with a medicine chest for basic first aid, plus a separate supply in case of a cholera outbreak containing a supply of opium pills, 'useful in looseness of the bowels', castor oil, spermaceti ointment, and Durham mustard for use as a poultice. Cholera was, in fact, a serious threat (there were major outbreaks in Scotland in 1832, 1848 and 1853) though, as the Commissioners pointed out, the isolation of most lights from the rest of the population slowed or stopped its spread to the keepers. All lights were also issued with a copy of the *Medical Directions for the Use of Lightkeepers*, which covered everything from correct use of leeches to the treatment of scurvy, poisoning, worms, smallpox and fainting.

Accompanying the standard-issue medicine chest was an extra box containing a dozen cases of essence of beef for making beef tea, two cases of arrowroot, three pots of blackcurrant, one bottle of brandy and three of port wine, all of which 'are intended to be used, and are to be used by the Principal or Assistant Lightkeeper and their families, but only in cases of sickness and of recovery from sickness, and when recommended by the Medical Attendant'. In theory, all lights were supposed to be dry. In practice, it was as impossible to banish all alcohol from the keepers' lives as it is in any other profession. At least one east coast light has an old smugglers' tunnel running straight through the rock beneath the tower to the nearest pub and Robert himself made a habit of presenting keepers' wives with several rusks and a bottle of sherry just before they gave birth. During this century, when the rules against drink became fiercer, several keepers used the lights as a treatment for alcoholism. They would drink themselves to stupefaction on leave and dry out on duty. As a self-administered detoxification programme, lighthouses might have been unconventional but they were peculiarly effective.

With the increase in lights came an increase in crimes. The General Order book filled with the babble of argument and disgrace, from the 'untidy condition of the assistant lightkeeper's bedding' to 'lightkeepers neglecting to affix requisite postage stamps to letters', from charges of 'intoxication' at

Skerryvore to the assistant keeper at Start Point 'shaking his fist in the face of the Principal'. 'The Commissioners,' the Secretary to the NLB Alex Cuningham wrote in 1866, 'at all times entertain a jealous anxiety for the respectability of the Lightkeepers as a body, and the Commissioners will visit with their severest censure any individual Lightkeeper whose acts tend to cast a slur upon or raise the voice of scandal against the general body of Lightkeepers.' The worst crime of all was to fall asleep while on duty. Those who did so even for two minutes could expect instant dismissal from the service. If they didn't inform on themselves, they could usually be confident that their colleagues would. To ensure that all keepers were properly wakeful for their spells on duty, the Commissioners insisted that they should not exert themselves unduly before taking the watch. Long walks or heavy exercise were forbidden and once in the lightroom, the keepers were banned from reading, writing or doing any other form of work. For four hours, the keepers were expected to watch the light, wind the clockwork and stare out to sea. Any other activity, however earnest or self-improving, was a distraction. 'The lightkeeper on duty,' said the Instructions to Keepers sternly, 'shall, *at his peril*, remain on guard till he is relieved by the Light-keeper in person who has the next watch.'

Robert in particular was disinclined to be generous to keepers abandoning their post even for more pressing reasons. In January 1813, he received a frantic letter from Andrew Darling, principal keeper at Pladda. A cargo ship had been wrecked on Arran and the crew had arrived at his light looking for shelter and help. Over a month later, they were still there, and were eating him out of house and home. 'Sir,' his letter read, 'Since the 12th of December the *Peggy & Betty's* crew have been living on me depending on Mr Whiteside (the captain) to send money for their relief and my good and use. Sir you cannot but see how I am served my provisions is all gone and has nothing to purchase more for my self and does not know what to do.

Therefore Sir for Godsake do some thing for me as I cannot do anything for myself from this moment they are owing me twenty two pounds sterling for board and Cash for the captain.' Robert's reply was unsympathetic. He was sending someone over from Dumbarton with money and a message to the crew, 'who from their stupid or criminal conduct in losing the ship do not appear to be worth house-room'. He told Darling that the incident would be investigated by the Commissioners, and furthermore that, 'either your lighthouse must have been faulty or his [Whiteside's] course have been strangely wrong'. Despite his occasional fits of goodwill, Robert could rarely be relied on to provide consolation. The keeper at the Mull of Kintyre wrote to Robert in 1820, nervously complaining that he had no transport or provisions since the lighthouse horse had fallen off the nearby cliff-edge into the sea and been killed. Robert replied grumpily that it would serve him right if he went hungry; he should have fenced off the cliffs to prevent just such an accident happening.

Arguments were usually a consequence of boredom, since much of the keepers' existence was spent cramped up with a colleague idly waiting for the next watch. Locked up together, many keepers developed mortal grudges against each other. At one English light, the assistant keeper's wife started a lengthy and bitter fight with the principal's wife because one had a doorstep to her cottage and the other didn't. Little Ross light, which guards the entrance to the Solway Firth, became infamous for the murder in 1960 of an occasional keeper by the assistant keeper. But when something serious did happen, the keepers would usually drop their quarrels and respond. On particularly dangerous stretches of sea, the keepers had to keep a regular watch for wrecks and groundings. Ships entering the Pentland Firth or the Strait of Corrievreckan could easily be caught by the current and spun out on top of a nearby rock. Yet there was little the keepers could do to stop such accidents happening. The NLB could have lit the coast like Christmas and still not been able to prevent the errors of a novice captain or a neglectful crew. Before the widespread use of radio, the keepers' only method of averting disaster was to signal to boats in trouble with flags or fire off warning flares.

Perhaps the oddest aspect of their job was that they were never specifically asked to save lives. The lights and their keepers were there as warnings, but those keepers were theoretically expected to do no more than to watch from the lantern windows as a ship ran aground on the rocks below. The original law of 1786 authorising the construction of the first four lights in Scotland never explicitly cited the lights as life-saving devices. The Act proposed only to build necessary beacons for 'the Security of Navigation and the Fisheries'. A later Act appointing the Commissioners allowed them to do whatever 'they shall think necessary or convenient for making, erecting, preserving and improving the said Light-houses and Works'. Even the NLB's motto is unspecific, *In Salutem Omnium*, 'for the safety of all'. It is a curious omission. The British Coastguard by contrast has as its primary aim 'to minimise loss of life amongst seafarers and coastal users', the RNLI states that it 'exists to save lives at sea', and even Trinity House in theory still adheres to its original poetic licence to 'succour from the dangers of the sea all who are beset upon the coasts of England, to feed them when ahungered and athirst, to bind up their wounds and to build and light proper beacons for the guidance of mariners'.

The Scots Commissioners, as fixedly prudent then as now, set themselves less high-flown goals. As they considered it, the point was to build the lights, fuel their burning, keep the keepers and collect the money, not to squander time and worry on the purpose of those lights. The existence of their sea-towers saved lives, albeit passively, and the Commissioners were never required to do more than build and maintain those towers in the ways they saw fit. In fact, the keepers were actively discouraged from helping victims of shipwreck, since to do so would have meant leaving the light and therefore neglecting their duties. Even now, sporadic acts of courage or selflessness by individual keepers – helping ships avoid trouble, personally warning of dangers, and occasionally risking their own lives to save sailors in distress – run contrary to the letter of the NLB's rules. The list of keepers' duties, which by 1850 ran to thirty-seven separate rulings, never included instructions on what to do in the event of shipwreck near their light. The only direction they were given was merely to 'take notice of' any shipwreck in the vicinity of their light and to maintain the Wreck Book. They could log a ship in distress, in other words, but they couldn't help save it. It is a wry little oddity that the Stevensons, building their lights to warn or save or fix position, never openly allowed their staff to do the same.

In practice, however, the keepers did not just stand idly by. The NLB's remit might not have directed anything more than maintaining the lights, but the keepers themselves adapted their role to the local realities. Dispensing weather reports to fisher-men, warning sailors of dangers or assisting ships that foundered were not part of their official duties, but many staff – including, tacitly, the Commissioners, would have questioned the notion that their only job was to watch and keep. Throughout the history of the NLB, there have been humbling acts of courage by individual keepers, from the rescue of twelve of the crew of the *Vicksburgh* by the Pentland Skerries keepers in 1884 to the stormy journey made by Robert Macauley, keeper at Fair Isle North, to help the keepers at the South light after a bombing raid during 1941. The General Order book, which mainly con-tained a shocked litany of keepers' crimes, also recorded several individual examples of heroism.

Whether saving tourists and trippers who had got into trouble, extinguishing the fires that occasionally broke out in the light-rooms or giving shelter to the victims of shipwreck, the keepers did not have a history of passivity. Medals from the Royal Humane Society were presented to a number of the keepers. One typical example was the medal presented to

William Davidson, principal lightkeeper at Tarbetness, for helping to save four out of five of the crew of a Norwegian schooner. As another of the rescuers reported, 'after the vessel struck She began to break up, when the crew jumped into the Sea amongst the wreckage and floating timbers, from which they were rescued with the greatest difficulty and danger, Mr Davidson and Mr McDonald being washed off their feet several times.' The Commissioners duly took note of Davidson's humanity and gallantry and were 'much gratified to know that a Lightkeeper in their Service has been judged worthy of so honourable a badge of merit'.

The unhappiest point in the keepers' history was not caused by shipwreck or storm but the disappearance of three keepers on the Flannan Isles, an incident that became the NLB's own unsolved *Marie Celeste*. The Flannan Isles are a remote cluster of rocks somewhere to the west of nowhere, populated mainly by sea birds and passing gales. Construction of a lighthouse on the largest rock began in 1896 under the supervision of David A. Stevenson. Like Dhu Heartach, the main rock was an immense sheer lump of sea-beaten reef without sheltered creeks or landing places. All building materials and provisions had to be hauled up the 150-foot cliff faces directly from the boats below with the full fetch of the Atlantic soaring straight over the summit. By 1899 the light had been finished, lit and staffed, but within a year an urgent report reached the Commissioners. The captain of the lighthouse ship the *Hesperus* had arrived at the light to make the regular relief on Boxing Day. As was usual, he hoisted the flag on the ship and waited for the answering signal from the light. When none came, he sounded the ship's siren, and then fired off one of the rockets to attract the keeper's attention. Again, there was no reaction from the lighthouse.

Later that evening, he telegrammed George Street. 'A dreadful accident has happened at the Flannans,' he wrote. 'The three keepers, Ducat, Marshall and the Occasional have disappeared from the island. On our arrival there this afternoon no sign of life was to be seen on the island. Fired a rocket, but, as no response was made, managed to land Moore (the relief keeper), who went up to the station but found no keepers there. The clocks were stopped and the other signs indicated that the accident must have happened about a week ago.' When Moore had reached the light, he had found no sign of life. The kitchen door was open, the fire laid but unlit, the beds empty and crumpled. The light itself was in perfect working order, the fuel supply full and the lamp freshly cleaned. On their return the next day, the same small group searched the island from end to end, still finding no sign of life. The log books had been maintained until a week previously, noting several days' worth of unexceptionable weather. Plates and cutlery in the kitchen had all been cleaned and left to dry, and the keepers' oilskins and seaboots were missing from the passageway. The only evidence of anything amiss was near the landing place to the light. By the cliff edge, where equipment for landing supplies would usually be stored, the party discovered several of the ropes and a box containing landing tackle missing and the iron railings guarding the footpath bent out of shape or wrenched from their foundations. A block of stone weighing at least a ton had somehow fallen several feet from the top of the cliff to the landing stage and a lifebuoy stored 110 feet above sea level had disappeared completely.

No absolute conclusion has ever been reached about the fate of the Flannan Isles keepers. The incident became the subject of fascinated national speculation and the source of several scandalous rumours. It was suggested that one of the keepers was an alcoholic who had pushed the other two over the cliff edge while drunk, that the three were involved in a soured love-affair, and even that one of the keepers had got God, and dragged his colleagues into the sea in a fit of religious contrition. Later rumours suggested all three had been abducted by extraterrestrials. The writer Wilfred Gibson wrote a poem shortly afterwards, speculating that an ancient curse on the island had turned the three keepers into black, raven-like birds. His contribution did nothing to silence the rumours.

The NLB and those who found the deserted light, however, held to a more realistic theory. In his official investigation for the Commissioners, lighthouse Superintendent Robert Muirhead concluded that 'After a careful examination of the place, the railings, ropes etc, and weighing all the evidence which I could secure, I am of the opinion that the most likely explanation of the disappearance of the three men is that they had all gone down on the afternoon of Saturday 15th December to the proximity of the west landing, to secure the box with the mooring ropes etc. and that an unexpectedly large roller had come up on the island, and a large body of water going up higher than where they were and coming down upon them had swept them away with resistless force. I have considered and discussed the possibility of the men being blown away by the wind, but, as the wind was westerly, I am of the opinion notwithstanding its great force, that the more probable explanation is that they have been washed away as, had the wind caught them, it would from its direction, have blown them up the island and I feel certain they would have managed to throw themselves down before they had reached the summit or brow of the island.' Freak waves are not such an improbable explanation. As Thomas Stevenson's experiments on Shetland had proved, the seas around the west coast of Scotland could touch the summit of the highest cliff, break up twenty-ton rocks with a flick of salt water and roar unbroken over 200-foot heights. Even on relatively calm days, the pulse of swell and tide produces occasional rogue waves bearing down on the surrounding land. Many of the last generation of keepers testify to alarming encounters with freak waves. Perhaps the peculiarity of the Flannan Isles incident was not that the sea had reached 110 feet up to where they were standing, ripped the railings from their roots and seized the men standing there, but that not one of the three was able to save himself.

The tragedy affected the rest of the service profoundly. Those who had found the deserted light suffered from bouts of sickness and guilt, while later keepers became distinctly reluctant to take up their postings to the Flannans. But this was not the only posting that proved unpopular in the service. Each light inevitably gained its own favour or notoriety, and certain postings became the NLB's equivalent of exile to Siberia. Of the rock lights, Skerryvore was favoured, since, for a tower, it was comparatively roomy and even at high tide allowed brief constricted walks around the reef. During heavy storms, though, the light would sway and flex in the wind. With the roaring water and the deafening isolation the keepers felt much as if they were living among the topmost branches of a high tree, with the foundations creaking below, the sea banging on the windows and the ghostly flicker of white spray falling past the lantern room. Keepers were not generally a superstitious breed, but Skerryvore could terrify or exhilarate the most flat-minded of men.

The Bell Rock, where all keepers served an initial apprenticeship, still carried its old prestigious veneer, but had less opportunity for exercise. Muckle Flugga was fondly regarded, since it was comparatively spacious and gave a grandstand view of the most spectacular seas in Scotland. Dhu Heartach and Chicken Rock (built by Tom and David off the coast of the Isle of Man) were resented for being small, cramped and uncomfortable. In the worst lights, so the myth went, the keepers developed peglegs from eternally climbing in circles and curved spines from sleeping in circular bunks. Some apprentice keepers got to the lights and found they couldn't bear the remoteness. Bruce Brown, the last keeper at Duncansby Head, remembers 'The rock stations, a molehill becomes a mountain. There's been supers [trainees] on these places, and they went melancholy. There was one super at Dhu Heartach – they called it the Black Hole at one time – and he got so bad he got down on the grating, he was going to dive off and swim ashore.'

But the loneliness of a lightkeeper's life is a myth. After the 1820s no light ever had fewer than three attendants. Thus the problem for new keepers was not in adjusting to solitude, but in living for a long time in a small room with a stranger. Too much of a liking for one's own company was, in fact, actively discouraged. A troubled correspondence took place in 1857 between George Street and the principal keeper William Primrose at the Calf of Man light, whose assistant keeper John Alexander, 'a very treacherous and dangerous man to be at large', was showing alarming signs of 'mental excitement'. Alex Cuningham, Secretary of the NLB, wrote worriedly to a local doctor, 'He has all along manifested in his conduct towards the other lightkeepers a misanthropic tendency. You will of course keep in view that a Lightkeeper has nightly several hours of solitary watching, and the serious consequences which might result from a momentary ebullition of mental derangement.' Having examined the patient, Dr Underwood wrote back, perplexed. Alexander was 'extremely eccentric', and the doctor himself noted that he took 'long walks into the country in search of a wife with money. He says she must have at least £2,000 and be able to perform well on the Piano and speak French fluently.' A few days previously, Alexander claimed that a large poker had fallen on his head, though the other keepers could find no sign of injury. Mrs Primrose, however, had become so worried by the threat of violence that she had become quite hysterical. Underwood judged that 'Alexander is labouring under a peculiar morbid state of mind, and I think it advisable he should appear before the Board.' Alexander was sacked and returned to his home town of Wick. It is not known whether he ever recovered, or ever found his musical wife.

The high point of the lighthouse service was the paraffin age, the eighty or so years between 1870 and 1950 when the majority of lights used paraffin rather than oil or electricity for fuel. The engineering had reached its technical zenith, the fuel burned cleanly and without interference, the lenses had been refined to a point of near-perfection, and the keepers had plenty to occupy them. Routine was stable but satisfying. Each keeper took a four-hour watch and during that time was responsible for ensuring that the light displayed the right character, that the fuel tanks were full and were feeding through correctly, that the winding mechanism (similar to that of a vast grandfather clock) was turning the light at the right speed, that the log books (containing weather, barometric pressure, wind speed and cloud base) were maintained and that nothing untoward was happening out at sea. The immense lenses would be polished early in the day, and the slim mercury bath on which they rested checked to ensure that nothing obstructed the smooth movement of the lenses. The machinery took half an hour or so to crank and ran down every two hours. If the keeper wanted to communicate with his colleagues, he used the mouthpiece in the lightroom. The only remaining example of the paraffin mechanism is at the first of the Northern Lights, Kinnaird Head, now a museum. Up in the lightroom, the lens glitters and shifts while the long chains of the winding mechanism turn without sound. The whole apparatus of the lens, with its brasswork and minutely angled prisms, weighs well over three tons, but one gentle push sends it turning as smoothly as cream in a churn.

Switches and electronic circuitry might be simpler but they scarcely compare to the polish-and-clockwork of the paraffin years. But with the advent of reliable electric lighting, there was no longer a need for fuel-burning lamps, however aesthetically pleasing they might have been. Electricity was capable of producing a beam of well over three million candlepower, could be increased or decreased at will, and dispensed almost entirely with the need for the immense storage tanks and winding mechanisms. Most produced their own power from diesel generators (mains electricity being both too unreliable and too awkward to be useful). The new lenses were less than a quarter of the size of the old and have decreased in size steadily ever since, while the elegant argand lamps have given way to a light only slightly larger than a domestic lightbulb. Electricity also possessed another vast advantage over oil or paraffin: it could be operated automatically. The need to have keepers perpetually on watch disappeared; the keepers merely had to press a switch on the wall, make the occasional check and leave the light to its own devices. And so, beginning in the 1960s, the NLB began automating the lights. An average of three lights a year would be adapted, and with them would go another huddle of keepers returning to civilian life. The NLB calls the process demanning; the keepers call it closure.

For the keepers, bred into fastidiousness over many generations, it is the mess of automation that is perhaps hardest to take. After the confusion and the bittersweet departure, the garden starts to silt with weeds, the paint cracks around the edges, the gutters drip, the rooms reek of damp, and the sea starts stealthily to repossess its old haunts. The major lighthouses were built to be occupied. Exposed to the elements, they need constant maintenance and regular check-ups. Though the NLB hopes that the costs will eventually come down, at present the price of keeping the automatic lights working efficiently is as much, if not more, than the price of employing three keepers. There is also something more elusive in their favour; something that, once lost, cannot be regained. As one keeper put it, 'The human presence is still paramount. It must be. When you see people who are lifesavers going, that's terrible.' His main worry about automation is for those who needed the lights in the first place, the mariners. 'Everyone knows that relying on GPS and satellite technology alone is complete rubbish,' he says. 'What happens when it fails? When there's no one to report on boats in danger, or give out local weather conditions?' The last generation of lighthouse keepers recognise the logic of automation, they just mourn its consequences. And, inevitably, they see the flaws in a system dependent on technology. As another keeper noted dourly, 'Computers don't get drunk, don't go insane, don't fall asleep. They just break down sometimes.' But, as Captain James Taylor, currently the NLB's Chief Executive, points out, the Stevensons would have approved of automation. If Robert could have built the lights to run automatically, he probably would have done so, considering his pride in machinery and his exasperation with the keepers. Given the debate over the loss of the keepers and the need for lights, it comes as a surprise to discover that the NLB does still build lights. The most recent examples, all built to guide the oil tankers sailing up the Minch on their way to the Sullom Voe terminal, are made of aluminium and just as sturdy as the first lights 210 years ago. None, needless to say, have ever been manned.

Two hundred years of professional history certainly made the keepers a singular breed of men. The necessary qualities for keeping were not the sort easily itemised on a job description. A love of details, an affection for the endless repetition of small tasks, a long-learned understanding of water, wind and tide, a habit of mind able to cope with dull days and ferocious nights. Bruce Brown, the last keeper at Duncansby Head, regarded patience as a prerequisite. 'There must be something in you. It's got to be there. Your outlook on life, it's all got to be thought about. Nothing bothers you; you just get that used to it that you just let it go, let it flow along. There's the rough, the smooth, the bad, the awful and the worst, and you just take it all.' Angus Hutchison, the man with the dubious honour of being The Last Keeper in Scotland (as principal at Fair Isle South), believes that good humour, equanimity, and a thorough understanding of human fallibility were the most important qualities. 'You need patience, a ready recognition of your own faults as well of those in the people around you, you need to make allowances. Most of us became very good at reading people's shortcomings and strengths.'

Many of the Scottish keepers had one particular quality in common, a vast and endearing capacity for understatement. Two-hundred-foot waves and winds strong enough to uproot fully grown trees were regarded as 'a bit rough out there tonight', a swaying tower light as 'an interesting experience', and a force-11 gale as ' a bit of a breeze'. Not everyone would relish a posting to the Flannan Isles, or consider themselves fortunate to watch 100-mph storm-force winds. Donald Michael, the last keeper at the Butt of Lewis, climbed to the top of the tower with his wife during the final gale of his tenure, just to watch. They couldn't hear themselves speak, so they just stood listening to the wind. 'It was an experience,' he says lightly. 'It was definitely something, watching that sea.' Most lights were fitted with railings around the cottages to prevent the keepers being hurled off the cliff edges by the wind. The men themselves deflect strangers' astonishment with a modest smile and a self-deprecating aside. The fables are left to spectators and trippers. Keepers do not feel the need for tall tales.

If you take the road north, go to the outermost point till the land stops and there is nothing but the broad dark horizon. Then go a little further. It doesn't particularly matter which road or which corner of Scotland you choose. Sule Skerry, Esha Ness, Ushenish, Skurdy Ness, Hyskeir, Auskerry, Ornsay, Muckle Flugga, Ruvaal, Skerryvore. Somewhere out there past the back of beyond will be a neat white wall, a few wind-scoured cottages and a tower. Seen from above, they look like punctuation marks between land and sea; a ragged grammar of full stops marking the end of Britain. Or, if you remember that these light squares are now abandoned, perhaps they seem more like graveyards. All that stone and history and effort, you think, just for a lightbulb.

TEN

Epilogue

To those who use and need the sea today, the argument over the lighthouses is less about their history than their current purpose. The way in which we treat the ocean has changed profoundly since James Park first arrived to polish the reflectors at Kinnaird Head. There are no more pirates, press gangs or wreckers. There are no eager opportunists waiting by the shore for ships to die. Those who work on the sea have become a marginal part of Britain's life, still there but somehow inessential, associated with a long-ago time before fish came in packets and boats became hobbies. Britain makes its money by other means; the sea has become optional now. It is possible to trace the sea's steady demotion through the political hierarchy. Two centuries ago, the Lord High Admiral was second only to the monarch and the prime minister. Now, civilian sea-business is dealt with by the Department of Transport or by the less-than-glamorous fisheries minister, a position that does not even carry a Cabinet place. Maritime safety is now governed by a flotilla of separate organisations: HM Coastguard, the Marine Safety Agency, the Marine Pollution Control Unit, the Royal Yachting Association and the Royal National Lifeboat Institution. The navy has been reduced to a peacetime rump, the fishing boats are preoccupied with quota systems and netting regulations, and the only criminals left are scraping a living from quota-jumping 'black fish'. Then Thomas Smith began his work, the sea was administered by the Admiralty, the criminals and the professionals. Now it has a civil service all of its own.

When the first lights were built around the British coastline, they were the only safety measures in existence. Of the 100,000 or so people employed on the sea in the 1750s, between thirty and forty per cent would not have survived to see old age. Those who escaped death through disease, ill-treatment or hardship had little hope of surviving shipwreck. The only chances of salvation in case of disaster were luck or land. Once down in the frozen Atlantic waters, they were as good as dead, and they knew as much. But alongside the slow development of the lighthouse services came other changes. The most significant nineteenth-century improvement in lifesaving was the foundation of the Royal National Lifeboat Institution. In 1823, Manby's old rival William Hillary published an impassioned pamphlet arguing the need for a regularised life-saving service. In his *Appeal to the British Nation of the Humanity and Policy of forming a National Institution for the Preservation of Lives and Property from Shipwreck*, he demanded to know why, 'In the nineteenth century, surrounded by every improvement and institution which the benevolent can suggest, or the art of man accomplish for the mitigation or prevention of human ills, will it for a moment be capable of belief, that there does not, in all our great and generous land, exist one National Institution which has for its direct object the rescue of human life from shipwreck?'

Hillary's pleas touched a public nerve. By the time of his death in 1847, he had personally helped to save over 300 lives, but, echoing Manby's example, he too died in extreme poverty, while the institution he founded languished. The RNLI, which now has a national volunteer crew of more than 4,400 and a record of saving more than 131,000 lives, came from uncertain beginnings. Until 1854, only four lifeboat stations were established in England and none at all in Scotland. The country had other things on its mind, and it was not until lifesaving was glamourised by Grace Darling's example that the public were sufficiently roused to support a full-time charitable organisation.

In part, they were galvanised by the events of a few fatal years. In 1851, the Board of Trade first started compiling figures for deaths at sea around the British coastline. Their findings made sobering reading. In that year, 428 people had been killed in one disaster alone, and in 1852, 1,115 vessels were wrecked and 900 lives lost. The figures did not include British shipping in foreign waters, which would probably have doubled the numbers.

Lifeboats were one thing; legislation was another. It took almost a decade of agitation before Samuel Plimsoll managed to push through his bill on load lines and finally sink the infamous coffin ships. The delay was largely due to the objections of shipowners, who regarded the measures as uneconomic and lobbied Parliament to drop the subject. Plimsoll finally became so exasperated by their flannellings over the measure that he lost his temper in the House of Commons and shook his fist at the Speaker. His outburst had cheering results. Public opinion swung his way and the Merchant Shipping Act of 1876 finally introduced the Plimsoll Mark. But not all of the delays were the fault of callous shipmasters or indifferent government. It took most of the nineteenth century before the argument between steam and sail was finally settled. During that time, the design and tonnage of steam ships evolved with bewildering speed. Shipping changed so rapidly that legislation could not keep up. It would have needed a separate Act for every category of vessel to allow for all eventualities. The casualties, meanwhile, continued to increase. During the 1880s, an average of 3,000 people died every year on merchant ships alone, and it was not until the turn of the century that the figures began to decline.

In 1890, a parliamentary committee considered legislation on bulkheads (vertical watertight divisions spaced down the hull sealing one part of a ship off from the next) but, despite the *Titanic*'s grim example, even this change was not introduced in all large ships until 1929. A further Shipping Act in 1890 also made it compulsory for all British ships to carry lifeboats, lifebelts, lifejackets (then made of cork), 'and other appliances for saving life at sea as are best adapted for the safety of the crew and passengers'. The Act was vague about provision and numbers, and neglected entirely the vexed question of ships above 10,000 tons. In practice, therefore, many shipowners skimped on the expenses, filled the boats with cargo or made access to safety equipment awkward if not downright impossible. Arguments still flourish about lifeboat provision on the *Titanic*, which at 60,250 tons was outside the terms of the legislation. Though there were twenty lifeboats, with provision for 1,178 passengers (23 per cent over the legal requirement) the ship was actually carrying around 2,200 passengers. After the disaster, some legislation was promptly enforced. All passenger or cargo-carrying vessels, of whatever tonnage, had to have a lifeboat place for every passenger, safety drill had to be regularly rehearsed, wirelesses monitored at all times, and all ships built double-hulled.

The greatest twentieth-century aid to saving lives was not initially the most obvious. By the turn of the century, wireless sets allowing communication between ship and shore were being developed. One of the first sets was fitted on the East Goodwin light vessel, allowing it to make contact with the North Foreland light. It proved its worth within a year when the ship was rammed. With the radio, the crew of the sinking ship were able to communicate with the shore and to send for help. By the outbreak of the First World War, many larger ships were fitted with wireless sets, although their initial range was limited and their signal fuzzy. But radio's arrival had broached the last great distance. Now, when a ship left port, it was no longer adrift on a wide wild sea, but linked to land, however ethereally, through the airwaves. Crews could receive and transmit weather reports, ice warnings, distress signals, jokes, insults and instructions. They were no longer sailing into the great unknown.

Progress in the past eighty years has been equally swift. Any ship that gets into trouble in British waters now has a recognised and well-established procedure to follow. A boat in danger within sight of the coastline would radio through a Mayday call, picked up either by the coastguard or by other nearby vessels. The coastguard then decides whether to send help and in what form (helicopter, speedboat or tug), or whether the problem could be better solved by instructing any other boats in the vicinity to assist. All vessels must, by law, note and act on a Mayday call and obey the coastguard's instructions to lend assistance. There are other emergency measures. Electronic beacons (EPIRB) that trigger on contact with water, sealed self-righting life-rafts, parachute rockets, radar reflectors and mobile phones. And, as of 1999, all ships should be fitted with GMDSS (the Global Maritime Distress and Safety System, an offshoot of the Global Positioning System, or GPS) which sends Mayday calls electronically, so overriding even the need for personal radio contact with the coastguard. None of the methods are completely fail-safe; none of them can guarantee to save lives, but collectively they do at least mean that anyone in trouble at sea has a far greater chance of surviving than sailors a century ago. These procedures, and their accompanying bureaucracy, have made the sea a safer place. So has its marginalisation; fewer people use it, those uses have changed, and those who 'follow the sea' theoretically do so in stronger, sturdier boats.

Not that any safety measure, however sophisticated, can override the waywardness of human nature. Insurance companies now recognise a phenomenon they have loosely termed Volvo Syndrome. The safer cars, boats or planes become, the more risks people take. Those who buy a Volvo are inclined to drive faster, rely more on the brakes and behave more aggressively than they probably would if they were driving something less impenetrable. The same impulse applies to boats. Swaddled in the comfortable technology of satellite systems, radar positioning, faxed forecasts and survival suits, sailors, particularly amateur ones, behave more recklessly than they would have done fifty years ago. In the last ten years, the number of drownings in British waters have remained stable at around 250 per year, but the number of incidents involving the coastguard, RAF or RNLI have more than doubled from 5,300 in 1986 to 11,300 in 1996. Likewise, the majority of accidents at sea do not involve merchant ships or fishing vessels, but 'leisure users', pleasure trippers who sail into unpleasant waters. The coastguard offers several explanations for the rise, such as a spate of warm summers, cheaper technology, and a growing number of sailors beguiled by quick charter deals. As they point out with some exasperation, these new fair-weather sailors are not always competent to take charge of a boat. Mayday calls (which are only supposed to be made when there is 'grave and imminent danger' to life or vessel) come in from crews who have run out of fuel or are feeling seasick. Even the lighthouses have unwittingly contributed to this sense of false security. For a long while, the English side of the Channel was better lit than the French side. Because of this, much of the passing traffic hugged the northern shoreline and, with so many vessels travelling such a narrow path, the number of collisions increased. Any benefit the lights had brought was therefore abruptly cancelled out; for every advance, there was also a retreat.

Besides, no instrument, however reliable or well-maintained, is infallible. If the microchips fail or the battery runs down (and about a third of all incidents reported to the coastguard do involve mechanical failure of one form or another) all sailors have to fall back on the skills of using compasses, stars, paper charts. The old techniques, like sailing with celestial navigation, or with sextants and chronometers, are still taught by professional organisations (the Royal and Merchant navies, the Royal Yachting Association). But, for amateur mariners sailing in inshore waters, there is in theory no reason why they should do more than study their GPS. With computers to fall back on, fewer people are learning their sums with quite the same rigour as in the past. As the law currently stands, there is nothing to stop anyone, however expert or incompetent, sailing the seven corners of the globe with less experience than it takes to paddle a canoe.

As of July 1998, all those who use the sea for work (fishermen, yachtsmen hiring out their boats, merchant skippers) have to have some form of qualification to sail in British waters. Amateur sailors need no qualification. There is no nautical equivalent of the compulsory driving test for cars. In the US, it is illegal to go to sea without insurance, and to obtain insurance, sailors need qualifications to demonstrate their ability. In Britain, the only check is on the use of a VHF radio. It is illegal to operate a radio without producing a Ship Radio Licence for which the owner needs a certificate of competence. Even then, the radio can still be operated by unqualified sailors under the owner's instruction and there is still no law forcing owners to possess a radio. Considering the rise in amateur sailing in the last two decades, the RYA points out that the general standard of sailing is remarkably high. They too are reluctant to insist on all sailors obtaining qualifications, since, aside from the difficulties in policing it, they believe that a universal minimum qualification would actually lower the standard of seamanship in Britain rather than improving it. The debate goes round in circles. For every improvement, another flaw develops.

But if those who use and follow the sea have changed, the sea has not. Very few aspects of existence on solid ground apply offshore. Water makes everything mutable. Once out of sight of the coast, life works to an alternative schedule, dictated by tides, winds, currents and stars rather than the fixed rotations of a clock. Practices are different, thinking is different, history is different. Even the geography is different. A hazard that appears one day can disappear the next. Despite our unresting efforts to immobilise our surroundings, they will keep moving beneath us. Land slips, coasts erode, sand banks up. The sea shifts its own furniture constantly. The best efforts of mankind are only meddlings by comparison. The lighthouse at Dungeness, for instance, has had to be rebuilt four times since 1600 because the sea kept sliding away from it. The most recent version, built in 1904, is already 500 metres from the high-water mark. Even in the most accurate chartings, errors creep in, currents change and politics takes an interest. One example is GPS, which gives each user an electronic guide to their situation, speed and time. GPS has become so popular and so cheap that most boats, however large or small, now use it. In theory, the system is accurate up to ten metres. At present, it relies entirely on twenty-eight American satellites run by the US military. During the Gulf War, the signals were deliberately distorted by several metres to prevent Iraqi intelligence from gaining too close a hold on Allied positionings. The world's sailors, in other words, are reliant on the goodwill of the Pentagon. Until the current system is replaced by international commercial satellites, politics will go on playing the same games with the sea as it did in the days of empires and monsters.

The lighthouses also became part of our Lilliputian effort to pin down Gulliver. They were just one in a long chain of attempts to civilise the sea, to make it manageable and rational. In the hands of a Smeaton or a Stevenson, they became part of a new faith. They were built for a specific purpose, but they also became expressions of a bolder time when Britain still believed that sense and science could conquer all. When the first lighthouses were built around the Scottish coast, they fulfilled a more complex role than merely warning against submerged dangers or acting as useful daylight landmarks. They acted as guides by which the mariner could fix position, calculate distance or gain an accurate measurement of speed. They also fulfilled a symbolic purpose as signs of civilisation, order and constancy. The lights told the mariner not only where he was but what course to take, and, if he made the right calculations, how long it would take him to get there. They acted as the ocean's stopwatches and alarms, and they became proof that no one, however apparently at sea, was ever quite alone.

Now there are plenty who claim that the lights have outlived their role. The same remote-controlled microchips that check the weather, measure the wind-speed and switch on the lights have also made those lights redundant. Computers and satellite systems can guide a sailor round the world, plot his course, fix the longitude, pilot the boat and find a route to safety without ever resorting to something as sweetly antique as a lighted tower on a headland. The world has come full circle – from darkness to light and back again. But to most sailors the lights remain necessary pleasures. They have become devices of last resort, used, in the old fashioned manner, when other methods have failed. They may have been superseded by technology, but that technology does not take kindly to force-10 gales and flooding waves. If it collapses, as it often does, there must be other methods and older forms of guidance. Which is why the first thing that a sailor will see as he fumbles through the darkness towards Britain is still a beam sweeping over the water, shining out the same false dawn. The sea is a tameable thing, and the lights have made it safe.

BIBLIOGRAPHY
INDEX

BIBLIOGRAPHY

GENERAL

Allardyce & Hood, *At Scotland's Edge* (London: HarperCollins, 1986).

Bell, Ian, *RLS: Dreams of Exile* (London: Headline, 1992).

Benham, Hervey, *The Salvagers* (Colchester, Essex Newspapers, 1980).

Burton, Anthony, *The Rise and Fall of British Shipbuilding* (London: Constable, 1994).

Bull, J. W., *An Introduction to Safety at Sea* (Brown, Son & Ferguson, 1981).

Balfour, Graham, *The Life of RLS*, Vols 1–2 (London: Methuen, 1901).

Baynham, Henry, *Before the Mast: Naval Ratings of the Nineteenth Century* (London: Hutchinson, 1971).

Broadie, Alexander (ed.), *The Scottish Enlightenment: An Anthology* (Edinburgh: Canongate, 1997).

Castle, Colin M., *Shipping and Shipbuilding on the Clyde* (Murdoch Carberry, 1990).

Chitings, Anand C., *The Scottish Enlightenment* (Croom Helm, 1976).

Cockburn, Henry, *Memorials of His Time* (Edinburgh: A & C Black, 1874).

Cordingley, David, *Life Among the Pirates* (Warner Books, 1995).

Cunningham, Alison, *Travels on the Continent 1863* (Chatto & Windus, 1926).

Daiches, David, *Edinburgh* (London: Hamish Hamilton, 1978).

Defoe, Daniel, *A Tour Through the Whole Island of Great Britain* (London, Penguin, 1986).

Ferguson, David, *Shipwrecks of Orkney, Shetland and Pentland Firth* (David & Charles, 1988).

Ferguson, David, *Shipwrecks of North East Scotland 1444–1990* (Aberdeen University Press, 1991).

Ferguson, William, *Scotland 1689 to the Present* (Edinburgh: Mercat Press, 1994).

Gifford, McWilliam & Walker, *The Buildings of Scotland: Edinburgh* (London: Penguin, 1991).

Hague & Christie, *Lighthouses: Their Architecture, History and Archaeology* (Dyfed: Gomer Press, 1975).

Hope, Ronald, *A New History of British Shipping* (London: John Murray, 1990).

Huggett, Frank, *Life & Work at Sea* (London: Harrap, 1975).

Gibson, W. M., *Old Orkney Sea Yarns* (Orkney: Kirkwall Press, 1986).

Lockart, J. G., *Life of Sir Walter Scott* (London: A & C Black, 1893).

Lynch, Michael, *Scotland: A New History* (London: Pimlico, 1992).

McLynn, Frank, *RLS: A Biography* (London: Pimlico, 1993).

Mair, Craig, *A Star for Seamen* (London: John Murray, 1978).

Majdalany, Fred, *The Red Rocks of Eddystone* (London: Longmans, 1959).

Mehew, Ernest, *Selected Letters of Robert Louis Stevenson* (London: Yale University Press, 1997).

Morris, Ruth & Frank, *Scottish Harbours* (Alethea Press, 1983).

Mowat, Sue, *The Port of Leith: Its History and its People* (Edinburgh: John Donald, undated).

Munro, R. W., *Scottish Lighthouses* (Stornoway: Thule Press, 1979).

Nicholson, Christopher, *Rock Lighthouses of Britain* (Whittles Publishing, 1995).

Parker, Tony, *Lighthouse* (London: Eland, 1975).

Pepys, Samuel, *The Shorter Pepys*, ed., Robert Latham, (London: Penguin, 1993).

Phillips-Bart, Douglas, *A History of Seamanship* (Allen & Unwin, 1971).

Rediker, Marcus, *Between the Devil and the Deep Blue Sea* (Cambridge University Press, 1987).

Rolt, L. T. C., *Victorian Engineering* (London: Penguin, 1988).

Scott, P. H., *1707: The Union of Scotland and England* (Chambers, 1979).

Smiles, Samuel, *Lives of the Engineers – Smeaton and Rennie* (New York: Scribner's Sons, 1905).

Smout, T. C., *A Century of the Scottish People 1830–1950* (London: Fontana Press, 1987).

Smout, T. C., *A Century of the Scottish People 1560–1830* (London: Fontana Press, 1985).

Smout, T. C., *Scotland and the Sea* (Edinburgh: John Donald, 1992).

Straub, Hans, *A History of Civil Engineering* (Leonard Hill, 1952).

Sutton-Jones, Kenneth, *Pharos: The Lighthouse Yesterday Today and Tomorrow* (Salisbury: Michael Russell, 1985).

Taylor, E. G. R., *The Haven-Finding Art: A History of Navigation from Odysseus to Captain Cook* (Hollis & Carter, 1971).

Talbot, Frederick, *Lightships and Lighthouses* (J. B. Lippincott Co., 1913).

Thrower, Norman, J. W., *Maps & Civilisation* (Chicago University Press, 1996).

Uden & Cooper, *A Dictionary of British Ships and Seamen* (Allen Lane, 1980).

Walthew, Kenneth, *From Rock and Tempest* (Geoffrey Bles Ltd, 1971).

Warner, Oliver, *The Lifeboat Service* (London: Cassell & Co., 1974).

A list of archival material and its location.

NORTHERN LIGHTHOUSE BOARD

Minute Books (1786 onwards), Scottish Records Office.
General Order Books (1844 onwards), Scottish Records Office.
Annual Reports, Scottish Records Office.
Northern Lighthouse Journal, the magazine of the NLB, NLB.
Returns and Statutes on Lighthouses (George III onwards), Scottish Records Office.
Letterbooks (1896 onwards) Scottish Records Office
Notices to Mariners (1896 onwards) Scottish Records Office
Shipwreck Returns, Scottish Records Office
Bell Rock Bible, Scottish Records Office

STEVENSON FAMILY PAPERS (ACC. 10706)

Letterbooks (outgoing) 1799–1847, National Library of Scotland.
Letterbooks (incoming) 1786–1891, National Library of Scotland.

Publications and ms of the Stevensons
Robert Stevenson

Travelling Journal 1789–1804, National Library of Scotland ms.
Travelling Journal 1826–1842, National Library of Scotland ms.
Corrected copy of *Account of the Bell Rock Lighthouse* (Edinburgh: Constable, 1824).
Travelling letterbooks, 1826–1842, National Library of Scotland ms.
Notebooks: 1811, 1812–1813, 1811–1812, National Library of Scotland ms.
Cuttings books: 1823–1853, 1853–1901, National Library of Scotland ms.
English Lighthouse Tours 1801, 1813, 1818 (Thos. Nelson, 1946).
Reminiscences of Sir Walter Scott, Bt., National Library of Scotland ms.

Alan Stevenson

Notebooks of visits to Northern Lights 1844–1849, National Library of Scotland ms.
Biographical Sketch of the late Robert Stevenson, (Edinburgh: A & C Black, 1851).
Account of the Skerryvore Lighthouse (Edinburgh: A & C Black 1848).
The Ten Hymns of Synesius, Bishop of Cyrene AD410 (1865).
Encyclopaedia Brittanica, entry on Lighthouses, (Chambers, 1864(?)).
The British Pharos (W. Reid & Son, 1831).

David Stevenson

Diaries 1830–1837, National Library of Scotland ms.
Lighthouse Inspection notebook 1860, National Library of Scotland ms.
Notebook of North American Tour, *Sketch of the Civil Engineering of North America* (London, 1838).

Life of Robert Stevenson, Civil Engineer (Edinburgh, 1878).
Lighthouses, from 'Good Words', (Edinburgh: A & C Black, 1865).

D. Alan Stevenson

The World's Lighthouses before 1820, (London: Oxford University Press, 1959).
Improvements in Seamarks before 1940: An Assessment, AISM Bulletin, 1961.

Thomas Stevenson

Skerryvore Diary 1843, National Library of Scotland ms.
Journal of a voyage around the Northern Lights 1847, National Library of Scotland ms.
Experiments and correspondence on the force of waves 1844–1853, National Library of Scotland ms.
Notebooks 1863–1879, plus 5 undated notebooks, National Library of Scotland ms.
Miscellaneous papers, undated, National Library of Scotland ms.
Lighthouse Illumination (London, 1859).
The Design and Construction of Harbours (Edinburgh: A & C Black, 1864).

Robert Louis Stevenson

Memories and Portraits, Tusitala Edition 1887 (London: Wm Heineman).
Poems Vol 1, Tusitala Edition, 1923.
Further Memories Tusitala Edition, 1923.
Records of a Family of Engineers Tusitala Edition 1924.
The New Lighthouse on the Dhu Heartach Rock, Argyllshire (edited by Roger Swearingen), Silverado Edition, 1995.
Kidnapped Tusitala Edition 1886.
Edinburgh: Picturesque Notes (Seeley & Co., 1900).
Scott's Voyage in the Lighthouse Yacht (Edinburgh University Press, 1989).
Memoirs of an Islet (Edinburgh University Press, 1989).

INDEX

Qualifying terms in brackets after Stevenson names refer to their relationship to Robert Louis Stevenson